SDG13 – CLIMATE ACTION

CONCISE GUIDES TO THE UNITED NATIONS SUSTAINABLE DEVELOPMENT GOALS

Series Editors
Walter Leal Filho
World Sustainable Development Research and Transfer Centre, Hamburg University of Applied Sciences

Mark Mifsud
Centre for Environmental Education and Research, University of Malta

This series comprises 17 short books, each examining one of the UN Sustainable Development Goals (SDGs).

The series provides an integrated assessment of the SDGs from an economic, social, environmental and cultural perspective. Books in the series critically analyse and assess the SDGs from a multi-disciplinary and a multi-regional standpoint, with each title demonstrating innovation in theoretical and empirical analysis, methodology and application of the SDG concerned.

Titles in this series have a particular focus on the means to implement the SDGs, and each one includes a short introduction to the SDG in question along with a synopsis of their implications on the economic, social, environmental and cultural domains.

SDG13 – CLIMATE ACTION

Combating Climate Change and its Impacts

BY

FEDERICA DONI
University of Milano-Bicocca, Italy

ANDREA GASPERINI
Italian Association of Financial Analysts, Italy

JOÃO TORRES SOARES
Universidade de Passo Fundo, Brazil

United Kingdom – North America – Japan – India
Malaysia – China

Emerald Publishing Limited
Howard House, Wagon Lane, Bingley BD16 1WA, UK

First edition 2020

Copyright © 2020 Federica Doni, Andrea Gasperini, João Torres Soares.
Published under exclusive licence by Emerald Publishing Limited.

Reprints and permissions service
Contact: permissions@emeraldinsight.com

No part of this book may be reproduced, stored in a retrieval system, transmitted in any form or by any means electronic, mechanical, photocopying, recording or otherwise without either the prior written permission of the publisher or a licence permitting restricted copying issued in the UK by The Copyright Licensing Agency and in the USA by The Copyright Clearance Center. No responsibility is accepted for the accuracy of information contained in the text, illustrations or advertisements. The opinions expressed in these chapters are not necessarily those of the Author or the publisher.

British Library Cataloguing in Publication Data
A catalogue record for this book is available from the British Library

ISBN: 978-1-78756-918-8 (Print)
ISBN: 978-1-78756-915-7 (Online)
ISBN: 978-1-78756-917-1 (Epub)

Printed and bound by CPI Group (UK) Ltd, Croydon, CR0 4YY

ISOQAR certified
Management System,
awarded to Emerald
for adherence to
Environmental
standard
ISO 14001:2004.

Certificate Number 1985
ISO 14001

INVESTOR IN PEOPLE

CONTENTS

About the Authors vii

Acknowledgements ix

Executive Summary x

1. Introduction 1

2. What are the Global Sustainable Development Goals? 3
   2.1. Introduction 3
   2.2. The Implementation of Agenda 2030 and the SDGs: A Global and National Perspective 7
   2.3. Actions to Implement the SDGs 10
   2.4. The Calculation of the SDG Indicator at 'Country' Level 12

3. What is the SDG 13? 21
   3.1. SDG 13 and the Main Focus on Climate Change 21
   3.2. SDG 13 Targets and Indicators 22

4. Climate Change Trends: Understanding SDG 13 Local Agenda and Key Interactions Frameworks 31
   4.1. Climate Change Agreement Signed in December 2015 in Paris (COP21) 31
   4.2. Interactions Frameworks 32

5. Practical Tools and Mechanisms for SDG 13
   Implementation   37
   5.1. Setting the Local Agenda: Governments and
        Public Sector   37
   5.2. Localisation of the SDGs   40
   5.3. Inclusiveness and the Participatory Approach   45
   5.4. Private Sector and Business Context   46
   5.5. Planning for SDG 13 Implementation:
        Climate Risk Assessment   47
   5.6. Innovative Financial Mechanisms:
        Green and Sustainable Finance   56
   5.7. ESG and Sustainable Finance   61
   5.8. Mobilising Green and Sustainable Finance   69

6. Monitoring, Evaluation and Reporting on
   SDG 13 Implementation   73
   6.1. New Trends in Corporate Reporting   73
   6.2. From Voluntary to Compulsory Disclosures   74
   6.3. SDGs and Global Reporting Initiative   77
   6.4. SDGs and Integrated Reporting   82

7. Conclusions   85

Appendix   89
References   97
Index   117

ABOUT THE AUTHORS

Federica Doni, PhD, is an Associate Professor in Business Administration at the Department of Business and Law, University of Milano-Bicocca. She teaches financial accounting and business analysis, business administration and fiscal and business planning and consultancy. Her main research interests include international accounting, intangibles and intellectual capital, integrated and sustainability reporting, corporate governance, accounting history and business valuation. She attends as a speaker at many international conferences, such as EIASM workshops and the annual EAA congress. She is also a Member of the Italian Association of Financial Analysts and Consultants, British Accounting and Finance Association (UK), European Accounting Association, European Academy of Management, the Centre for Social & Environmental Accounting Research (University of St. Andrews, UK), <IR> Academic Network (IIRC, London, UK) and PRI Academic Network (London, UK).

Andrea Gasperini is a Chartered Accountant and an Auditor in Milan, AIAF Member (Italian Association for Financial Analysis) since 1993 and the Head of Sustainability and 'ESG Observatory' for AIAF, which deals with all topics about sustainable finance and a strong proponent that climate change impacts financial investment decisions. Since February 2019 he has been a Member of the European Lab Project Task Force on Climate-related Reporting in European Financial

Reporting Advisory Group (EFRAG). His professional commitments concern all matters related to sustainability, particularly the risks and opportunities connected to a transitioning to a low-carbon economy resilient to climate change[1]. His other expertise includes financial analysis of the methodology to identify and communicate the intangible assets and liabilities, the environmental, social and governance factors the GRI Sustainability Report and the Integrated Report according to the IIRC framework of the German Council Sustainability Code and the Sustainable Development Goals Agenda 2030 UN. He is the author and co-author of many books and a lot of essays, published in several management and economic journals about sustainable finance and intangible assets. He has been a keynote speaker at the university and conferences planned on these fields by national and international organisations.

João Torres Soares completed his MSc degree in Engineering at the Post-Graduate Programme in Civil and Environmental Engineering (concentration area of infrastructure and environment) from the University of Passo Fundo. He is a Postgraduate in Environmental Law and Economics at the European University Institute – Florence and École des Hautes Études en Sciences Sociales – Paris; and in Business Reception Areas Management at the University of Aveiro – Portugal. Working as a Consultant, he worked in business strategy and local and regional developments. He held management positions in the public sector and in private non-profit institutions. He is a Co-organiser of The Plankton Symposium's first and third editions and of the First World Environmental Education Congress.

ACKNOWLEDGEMENTS

Andrea Gasperini would like to thank the Italian Association for Financial Analysis (AIAF) for the collaboration over the years about research projects on intangible assets and the evolution of knowledge about the intellectual capital. More recently, special thanks is given for the support received for all initiatives on the topic of sustainability by the Working Groups on 'Legislative Decree 254/2016 and non-financial information' and 'Sustainability and ESG information' through the publication of the AIAF magazine and white papers. The last one is focussed on the issues of climate risks and the environmental, social and governance information to which there are some references for a broader study of the topics discussed in this book.

EXECUTIVE SUMMARY[2]

This book aims to explain the Sustainable Development Goals' (SDGs) approach for sustainable development by analysing in depth one of the most important sustainable goals, that is, SDG 13 'Combating climate change. Take urgent action to combat climate change and its consequences'. This goal constitutes one of the most effective actions to protect and preserve our planet from the consequences of climate change. The concept of 'sustainable development' that was defined by the World Commission on Environment and Development (UNWCED, 1987) should ensure prosperity and environmental protection without compromising the ability of future generations to meet their needs. It therefore combines economic development with social inclusion and environmental sustainability. The SDGs targets enable governments, companies and investors to monitor their own progress in achieving these goals, which include ending poverty, eliminating hunger, addressing gender equality and combating the effects of climate change. According to UN statistics, the change in global weather conditions is the major threat to sustainable development, such that 12 out of 17 SDGs require action on climate change to address the main subject of the goal, and Goal 13 is specifically dedicated to climate actions directed at reducing emissions and building climate resilience. The achievement of the objectives can be affected by the possibility of obtaining data from new sources that allow for more detailed and 'granular' information. To doing this, an interesting index, that is, an SDG ('unofficial SDG index') indicator was described and calculated at 'country' level. After a brief analysis on the SDGs approach, the main focus is on SDG 13 and climate change. To address climate change, countries adopted the Paris Agreement at the COP21 (December 2015) and are committed to achieve five targets and eight indicators related to the SDG 13. SDG 13 requires urgent action on climate change that is central to delivering sustainable development (UNDP, 2017). Tackling climate change through adaptation and mitigation can represent a great tool to drive sustainable development outcomes in the key areas of Agenda 2030. It is worth noting that climate change is 'a cross-cutting issue' to be addressed in order to achieve a successful implementation of all 17 SDGs. Regarding the specific actions it is important to adopt climate measures in three key action areas: that is, *climate change adaptation*, *zero-carbon development* and *scaled-up climate finance*. The agreement on climate change (COP21) by 196 states invited all countries to accelerate and intensify the actions and investments necessary for a sustainable future, low-carbon economy resilient to climate change.

Actions to achieve the SDG 13 targets have to take into account the interactions between SDG 13 and other SDGs through an overall consideration of synergies and potential connections among different sustainable goals. An important step is to consider and evaluate some practical tools and mechanisms for SDG 13 implementation, especially focussing on setting the local agenda emphasising the role played by governments and public sector. An interesting analysis has been carried out by the Climate Action Tracker and Climate Transparency on both the content of intended nationally determined contribution (*what governments propose to do*) and current policies (*what governments are actually doing*) on climate mitigation in G20 countries. From this perspective, it is becoming crucial to strengthen the accountability system in order to measure progress on sustainable development to complement gross domestic product. Another important factor is represented by some public management tools such as budgeting practices and procedures to foster greater civic participation and more inclusive societies. To stimulate awareness raising, it is not enough to inform citizens about the existence of the SDGs; it is also important to act on empowering them to participate in the achievement of the SDGs in their daily lives by adopting an open participatory approach. Public sector and governments can play a crucial role to the achievement of SDG 13 but also private organisations and companies can support countries in the transition to a low-carbon economy. Companies have to communicate – to investors, credit and sustainability rating agencies, insurance companies, credit institutions and other stakeholders – further standardised information on their exposure to climatic risks. For implementing the SDG 13, companies and private sector have to adopt a climate risks assessment, as a growing number of investors are aware that climate change may affect their financial performance, if they are not able to evaluate correctly the risks. To better identify the information required by investors, asset managers, financial analysts, lenders and insurance underwriters, and to adequately assess climate risks and opportunities, the Financial Stability Board (http://www.fsb.org/) has set up a working group called Task Force on Climate-related Financial Disclosures whose task is to promote voluntary and coherent financial communication about climate change (FSB, 2015a, 2015b). This initiative supported the *Financing Sustainable Growth Action Plan* adopted by the European Parliament and the European Council in March 2018. It represents a roadmap with specific measures and related deadlines with the aim of (i) redirecting capital flows towards sustainable investments in order to achieve sustainable and inclusive growth, (ii) managing financial risks arising from climate change, resource depletion, environmental degradation and social issues and (iii) promoting transparency and long-term vision of economic and financial activities (DB Climate Change Advisors, 2012).

This roadmap is part of a wider European initiative on sustainable development that placed environmental, social and governance (ESG) issues at the heart of the financial system to support the transformation of the European economy into a greener, more resilient and circular system. Clear indicators of ESG topics has therefore begun to be requested by institutional investors who increasingly take into account the integration of ESG metrics to jointly evaluate the value creation of companies and sustainability (Cotter & Naja, 2011). For interpreting sustainable finance, different tools can be considered such as *Exclusionary Screening*, *Best in Class Selection*, *Engagement and active ownership*, *Thematic Investing*, *Impact Investing and ESG Integration*. Not only financial industry but also business leaders should identify essential resources and related threats that have impact on running their businesses, while looking at and accounting for the environmental and social impacts of corporate activities. These practices are increasingly considered as strategic tools that allow CEOs and a company's management to learn about risks related to the external and internal environment, and to design tailor-made strategies for adapting to those potential threats and turn them into opportunities. Corporate reporting may represent an important tool to effectively communicate SDGs to external stakeholders by adopting different frameworks, such as Global Reporting Initiative standards and SDGs Compass Guide through some specific SDG 13 indicators. Another important challenge in corporate reporting and SDGs disclosure can be supported by integrated reporting (Gasperini & Doni, 2017c; IIRC, 2013). Finally, measures and actions to prevent and combat climate change represent an important challenge for all countries, at a global and national level.

NOTES

1. Appendix provides an overview of definitions of climate change (pp. 89–96).

2. This book is the result of collaborative analysis: Chapter 1 and Appendix are written by João Torres Soares; Section 2.1, Section 4.1, Sections 5.4–5.8 and Sections 6.1–6.3 are written by Andrea Gasperini; and Sections 2.1–2.3, Chapter 3, Section 4.2, Sections 5.1–5.3, Section 6.4 and Chapter 7 are written by Federica Doni.

1

INTRODUCTION

It is not easy to live in a world of rapid and profound change. The 17 Sustainable Development Goals (SDGs) appeared, on the edge of the fourth technological revolution, to help and guide mankind's efforts to avoid catastrophic scenarios for our own and future generations' living conditions. By attempting to gather and synchronise all stakeholders around common goals and measurable targets, we hope to achieve the best balance for the world by 2030. At the very start, all statistical agencies were invited to harmonise working methods to achieve a reliable and compatible information database. This has been accomplished; now it is time to establish alignments and a common language for all, including those, who in their own homes want to contribute positively to this major global effort. The first step, set out in this chapter, is to make SDG 13 – 'Take urgent action to combat climate change and its impact' – clear to all. With this in mind, the authors have desegregated the concepts and looked for definitions in different research areas in order to create concepts that are as cross-cutting as possible. Our other main challenge was to express these concepts in simple, current and attractive language so that they may be read and understood by all, young and old, from the more educated to those who rely solely on their good, and will help humankind in this enormously significant task.

2

WHAT ARE THE GLOBAL SUSTAINABLE DEVELOPMENT GOALS?

2.1. INTRODUCTION

This chapter aims to explain the SDGs' approach for sustainable development by analysing in depth one of the most important sustainable goals, that is, SDG 13 'Combating climate change. Take urgent action to combat climate change and its consequences'. This goal constitutes one of the most effective actions to protect and preserve our planet from the consequences of climate change.

In the report published in the Intergovernmental Panel on Climate Change (IPCC, 2013a, 2013b, 2014), a leading international organisation for the assessment of climate change, highlights that human influence on the climate is clear and recent anthropogenic emissions of greenhouse gas (GHG) are the highest in history. The urgent need to combat climate change is an increasing requirement for all countries in the world and can be supported by financial markets and institutions at a global level. The enormous pressure to stimulate financial mechanisms signals the importance to involve finance in accelerating the global process to SDGs and achievements.

In this context, the year 2015 was extraordinary for growing and widespread awareness about the events that preceded and followed the launch of the UN Agenda 2030, such as the publication of the Encyclical Letter *Laudato Si* by Pope Francis[1] (May 2015) and the Paris Agreement on climate change (Paris, December 2015). The attention from the financial system to environmental and social issues was stimulated by the Encyclical Letter 'on the care of the common home' published on 18 June 2015 which deals directly with the issues of the world of finance, its operations and practices, its relationship with the real economy and its impact in terms of social justice and environmental protection. The aim of this chapter is not to study and classify phenomena (even in moral terms) but to highlight the urgency of change and to push for action[2]. This Encyclical Letter highlights that an energy supply system based on fossil fuels is primarily responsible for global warming and, therefore, for climate change:

> *[...] the climate is a common good, belonging to all and meant for all. At the global level, it is a complex system linked to many of the essential conditions for human life. A very solid scientific consensus indicates that we are presently witnessing a disturbing warming of the climatic system. In recent decades, this warming has been accompanied by a constant rise in the sea level and, it would appear, by an increase of extreme weather events, even if a scientifically determinable cause cannot be assigned to each particular phenomenon.*
>
> *Humanity is stimulated to recognize the need for changes of lifestyle, production and consumption, in order to combat this warming or at least the human*

> *causes which produce or aggravate it. It is true that there are other factors (such as volcanic activity, variations in the earth's orbit and axis, the solar cycle), yet a number of scientific studies indicate that most global warming in recent decades is due to the great concentration of greenhouse gases (carbon dioxide, methane, nitrogen oxides and others) released mainly as a result of human activity.*
>
> *Concentrated in the atmosphere, these gases do not allow the warmth of the sun's rays reflected by the earth to be dispersed in space. The problem has been exacerbated by a model of development based on the intensive use of fossil fuels, which is at the heart of the worldwide energy system. Another determining factor has been an increase in changed uses of the soil, principally deforestation for agricultural purposes. (LS 23)*

The document emphasises the crucial role played by the field of finance in the transition to a low-carbon economy (Carbon Tracker Initiative, 2011, 2013; Carbon Tracker Initiative/Climate Disclosure Standards, 2016); it can support the development of initiatives in the field of renewable energy, proceeding step-by-step according to need to avoid imbalances and shock, as indicated by Pope Francis with respect to energy transition:

> *[…] we know that technology based on the use of highly polluting fossil fuels – especially coal, but also oil and, to a lesser degree, gas – needs to be replaced immediately. Until greater progress is made in developing widely accessible sources of renewable energy, it is legitimate to choose the lesser of two evils or to find short-term solutions. (LS 165)*

Hence, in September 2015, the world leaders of 193 countries met at the United Nations to approve 17 SDGs and 169 targets indicated in the UN paper *Transforming our world. The 2030 Agenda for sustainable development.* This document is aimed at promoting prosperity by the end of the year 2030 for everyone and a more sustainable path for our planet and economy.

Agenda 2030 states that:

> *[...] The Sustainable Development Goals and targets are integrated and indivisible, global in nature and universally applicable, taking into account different national realities, capacities and levels of development and respecting national policies and priorities. Targets are defined as aspirational and global, with each Government setting its own national targets guided by the global level of ambition but taking into account national circumstances. Each Government will also decide how these aspirational and global targets should be incorporated into national planning processes, policies and strategies. It is important to recognize the link between sustainable development and other relevant ongoing processes in the economic, social and environmental fields.*

In this context, the concept of *sustainable development* was defined by the World Commission on Environment and Development in a document entitled *Our Common Future* within its Brundtland report. This document maintained that sustainable development should ensure prosperity and environmental protection without compromising the ability of future generations to meet their needs (UNWCED, 1987). It therefore combines economic development with social inclusion and environmental sustainability. The SDG targets enable governments, companies and investors to monitor their own progress in achieving these goals, which include ending poverty, eliminating hunger, addressing gender equality and combating the effects of climate change. SDGs can be categorised and described in different ways. For example, in the private sector, the company Morgan Stanley Capital International (MSCI) provides one way of classifying the SDGs for investors by grouping them into five categories: basic needs, empowerment, climate change, natural capital and governance. This categorisation has been developed to provide a framework for evaluating whether companies' business models and revenues relate to these categories rather than the 17 individual goals. Each category is linked to issues that can identify the SDGs. For example, the category 'Basic needs' is linked to 'No poverty' (SDG 1), 'Zero hunger' (SDG 2), 'Good health & well-being' (SDG 3), 'Clean water & sanitation' (SDG 4) and 'Sustainable cities & communities' (SDG 5) (UN SDGs, MSCI ESG Research, www.msci.com).

2.2. THE IMPLEMENTATION OF AGENDA 2030 AND THE SDGs: A GLOBAL AND NATIONAL PERSPECTIVE

The SDGs are a continuation of the Millennium Development Goals (MDGs) published by the United Nations in 2001, setting eight initiatives to improve the world focussing on social goals by 2015 (https://www.undp.org/content/undp/en/home/sdgoverview/mdg_goals.htm). Through the MDGs, parties attending the Summit in Rio de Janeiro (1992) committed their nations to eradicate extreme poverty and hunger, achieve universal primary education, promote gender equality and empower women, reduce child mortality, improve maternal health, combat HIV/AIDS, malaria and other diseases, ensure environmental sustainability and develop a global partnership for development.

Moving from the MDGs to the SDGs has highlighted some issues about their implementation at a global and national level. Successful implementation of the SDGs requires an intensive effort of international cooperation and a global commitment by many actors from countries all over the world. From this perspective, the SDGs show the ambition to stimulate world leaders together with all stakeholders in taking integrated actions with economic, social and environmental dimensions. International and national policymakers are expected to set frameworks, which enable new and enhanced collaborative global partnership between all actors to achieve the goals by mobilising all resources available and reshaping modes of production, consumption and living. Climate change is affecting everyone and everything around the world. However, those most vulnerable and impacted are also usually the ones living in the poorest countries or in regions often exposed to climate-related hazards and natural disasters. These categories, in turn, have more difficulty in moving resources to build stronger economies and safer, healthier and more livable societies. According to UN statistics, the change in global weather conditions is the major threat to sustainable development, such that 12 out of 17 SDGs require action on climate change to address the main subject of the goal, and Goal 13 is specifically dedicated to climate actions directed at reducing emissions and building climate resilience. Implementing policies and strategic plans to adapt and, where possible, mitigate climate change-related effects is the main priority of most nations attending the UN Framework Convention on Climate Change (UNF-CCC). Parties were to present every five years their 'intended nationally determined contributions' (INDCs). These represent the efforts each country has put into pursuing long-term goals, into reducing national emissions and preparing communities for the impact of climate change (UNFCC, 2014).

COP22, held in Marrakech, and COP23, hosted in Bonn in November 2017 and for the first time presided over by a small island developing state (SIDS; in this case by the Presidency of Fiji), further reinforced the parties' commitment to taking urgent action across a number of social dimensions and, in particular, on climate change. The COP23 was the first summit after the election of the President of United States, Donald Trump, and his announcement to withdraw from the Paris Agreement. The shock and uncertainty that Trump's decision has provoked on the future of the Paris Agreement, on funds available for climate mitigation and future cooperation, especially among US allies and countries dependent on the United States has been dramatic. Trump's presidency has undermined the collaborative partnerships and relationships established since Rio Earth Summit. Furthermore, it has threatened the success of years of meetings and efforts. Nevertheless, federal states (with the autonomy to take action on climate change, e.g., California), cities and many companies in the United States are sensitive to climate change-related issues and are committing themselves by implementing adaptation plans, as well as building partnerships and demanding cleaner energy.

However, the rate of climate change, together with other indicators of global well-being, is alarming. Further measures should be implemented, in order to accelerate the transition to a low-carbon economy and to remedy, as soon as possible, those difficult conditions that worsen and exacerbate the impact of natural disasters, crises and food and water scarcities (Mpandeli et al., 2018). Some important global trends are crucial as they can challenge and further threaten the implementation of policies for sustainable development, as well as the achievement of the SDG targets. However, they can offer significant opportunities for strengthening multilateral partnership and global participation, also for improving and adapting technology and science in seeking solutions to global issues.

2.3. ACTIONS TO IMPLEMENT THE SDGs

The project has this essential aim: 'leave one no behind'. The main difficulty will therefore be its concrete realisation or, more specifically, moving 'from agreement to action'. In other words, it will be important to identify the operational and technical tools that can support different countries in achieving the various goals. Concrete action to support governments can be provided by the Organisation for Economic Co-operation and Development (OECD), as the United Nations have to monitor progress in relation to the objectives and verify the adoption of certain policies by states that are based on evidence-based decisions. In essence, the OECD has to provide innovative statistical methods and systems capable of capturing the radical change that the launch of the SDGs is bringing about in many countries. Through analysis of the various SDGs, it is possible to highlight that many of the goals are complex, interrelated and characterised by multiple issues. Hence, achieving the objectives may be conditioned by the possibility of obtaining data from new sources that allow for more detailed and 'granular' information.

To this end, a working group was set up between the end of 2015 and the beginning of 2016, joined by 28 countries and supported by some observers, including the OECD, which developed a global framework of indicators including around 230 new specific metrics designed for reaching all the goals and targets. Quantifying the SDGs can be perceived as a challenge, since no country is able to access all the data needed to calculate the various indicators. At a global level, progress achieved has to be closely monitored; official reports have to be drawn up and debates carried out in relation to the various policies have to be adopted. Furthermore, on thematic, regional and national levels, reports will have to be prepared, including follow-ups and reviews of the SDGs, and the OECD is collaborating in different countries to facilitate and strengthen the collection of data, in particular on environmental and governance issues. On a national level, it will then be important for each country to be able to implement and adapt the global objectives to the situation and objectives set at the local level. For this purpose, the OECD has to produce a document that serves as a preliminary assessment of each nation with regard to the various SDGs, also taking into account the economic situation and the responsibilities of the specific country.

The OECD is committed to developing new indicators in the areas of technology, income distribution, health, work, international investments and regional analysis. It is also essential to bear in mind that many of the SDGs targets stimulate certain behaviours by a particular nation that will inevitably have consequences on the performance of other countries. It is therefore necessary to assess the extent of these impacts: for example, this is possible through the world input–output tables that the OECD makes available to the various countries in order to control the 'transboundary' impacts of production and consumption in the OECD countries on CO_2 emissions and critical natural resources. By the same token, the OECD is also developing other metrics, such as, for example, the total official support for sustainable development capable of providing extensive information about natural resource flows. The creation of a statistical database able to describe the progress of countries in achieving the SDGs is a significant investment of resources and skills; it is producing a radical change in improving access to data collection and allowing better data collection, more detailed interpretation and clearer communication.

2.4. THE CALCULATION OF THE SDG INDICATOR AT 'COUNTRY' LEVEL

On 20 July 2016, two international organisations, the Bertelsmann Stiftung, a private foundation with social purposes, and the Sustainable Development Solutions Network (SDSN) organisation, promoted by the United Nations, presented an interesting document called the SDG Index & Dashboards to prepare operational solutions for sustainable development. A global report (July 2016), sent for public consultation, suggested an SDG ('unofficial SDG index') indicator described and calculated at 'country' level (Gasperini & Doni, 2016). This document is the result of the work of a team of experts, supported by the SDSN Secretariat, and attended by representatives from all over the world. The aim of this report was to provide a 'report card' in order to carry out a preliminary survey on the state of the various countries in terms of achieving the 17 SDGs and to guarantee accountability. The report calls on countries to proceed swiftly, albeit ambitiously, towards achieving these goals by implementing reforms, as immediate and global action is needed in the first years of implementing the new global agenda.

This is an indicator, for the time being unofficial, whose construction seems to be motivated by the need to combine three different crucial aspects for sustainable development, that is, economic development, inclusion of social aspects and environmental sustainability supported by good governance. The development of appropriate indicators and the collection of data can therefore transform the set of SDGs into a 'practical tool' for the solution of problems through the involvement of governments, academics, companies and communities. In the same way, the creation of a kind of 'report card' capable of recording progress achieved annually with the various SDGs is, therefore, a managerial tool for enacting transformations that allow the SDGs to be achieved by 2030. The development of an unofficial index of SDGs seeks a quick way of monitoring success in achieving the SDGs and in identifying the priorities for the 34 OECD countries regarding the various actions required and related to each goal. The document is basically divided into two parts: the first part considers the methodology used to develop the SDGs indicator and the second part describes the dashboards that can be used by individual countries. However, the calculation methods have some limitations which make these tools still unofficial and subject to revisions and improvements.

The SDG index refers to data for 2015 and allows countries to check progress and compare themselves with their peers or, in other words, with countries with similar income or in geographically similar areas. The indicator development methodology includes a series of indicators calculated in relation to each of the 17 goals for which the most recent available data have been collected. The unavoidable presence of missing values made it necessary to exclude some countries from the calculation of the indicator; some observations were also eliminated from the calculation as 'outliers'. Each indicator is ordered from the worst to the best value in order to create a 'best score'. In some cases, the best value corresponds to zero, as, for example, in the case of the 'no poverty' SDG, or 100% as in the education-related goal. There are then indicators for which it is not possible to ensure the issuing of a value that corresponds to the best level (e.g., infant mortality rate, etc.). This approach uses the five best values for each indicator and then proceeds to calculate the average value. If a country reaches a level higher than this average value, it is assigned the best value. Using a summary of the various indicators relating to the various SDGs, values are obtained for each country that are then summarised in a single index relative to a given country. These results reveal the data relating to indicators for each goal and summary indicators for each country: from a first assessment, the country with the highest overall score (84.5) is Sweden followed by Denmark and Norway. In the 2018 report, the same findings about the SDG index are confirmed. All countries in the top 20 are OECD countries. However, countries perform well on the Index score significantly below the maximum 100. Every country scores 'red' on at least one SDG in the Dashboards, for example, Italy shows four red scores related to SDGs 9, 12, 13, 14 and 17. Bertelsmann Stiftung and Sustainable Development Solutions Network, (2018, p. 18).

It is interesting to consider that, despite the scores, there is still a lot of progress to be made: for example, the shifting of the energy system from a high-carbon approach to a low-carbon one is essential in order to reach SDGs 7 and 13. Another aspect to consider is that some countries with a high income are far from achieving the SDGs; this is not surprising, as sustainable development includes the three pillars: economic, social and environmental supported by good governance. A high income at country level could be achieved thanks to conduct based on inequalities and not on properly sustainable environmental practices. Also, interesting is the comparison made between the SDGs' index scores and another indicator, the UNDP's Human Development Index: from the results it emerges that, although there is a high correlation between the two indicators for some countries, there is a substantial difference between the two indicators that shows, therefore, a misalignment between the two areas: the more limited one relative to human capital, and the broader one focussed on the SDGs. The second part of the document is dedicated to analysis of the SDGs dashboards created for each country: in essence it is a different way of representing the various SDGs indicators using a colour-based method ('colour-coded scheme'). The objectives are highlighted with the colours green, yellow or red, where the latter emphasises the need for radical changes in order for achievement of the goal to be likely. Green means that, for this indicator, the country is well on the way towards achieving a SDG and its objectives, or has (in some cases) already reached the threshold consistent with the implementation of the SDGs. The development of the dashboard involves the determination of four quantitative 'thresholds': the best and the worst scores (as indicated in the SDG index method); the threshold for reaching the goal; the threshold relative to an intermediate colour between red and yellow. In more details the green band is bounded by the maximum that can be achieved for each variable (i.e. the upper bound) and the threshold for achieving the SDG. There are three color bands ranging from yellow to orange and red that denote an increasing distance from SDG achievement. An overall colour rating is then generated for each goal that represents the 'minimum' or the lowest value of the 'colour rating' among the various indicators for a given SDG. Thus, for a given SDG, if a country registers red, even for just one indicator, and yellow for all the others, it is still marked red. This approach seeks to motivate the countries concerned to make radical changes to confront the situation.

From this analysis, we can highlight the following:

- The SDGs represent an 'action agenda' for rich and poor countries alike.

- Each OECD country must acknowledge red grading for different SDGs, which implies the need to adopt radical changes.

- The countries that are analysed show more than a third of the goals highlighted by red colour, where this means that at least one indicator has found a very low value.

- There are major changes related to climate change (SDG 13), ecosystem conservation (SDGs 14 and 15) and sustainable consumption and production (SDG 12).

- there is a substantial number of countries with red in relation to SDG 2; this means that the farming system is not sustainable or there is a high rate of obesity in the population and, therefore, of malnutrition.
- A high number of countries with problems with SDG 17 relating to the lack of financial resources to cope with the development of international cooperation.
- Significant difficulties reported by different countries in relation to SDG 8 for unemployment and low growth, and SDG 5 for deficits in relation to gender equality.

The report allows us, therefore, to have a global vision on the progress of the various SDGs and to obtain some indicators, even at the overall level, significant enough to be able to make comparisons between countries and the main geographical areas in the world.

However, four limitations are highlighted:

(1) The impossibility of monitoring and evaluating some SDGs through different national contexts as only a few SDGs allow an evaluation of cross-country effects and the benefits deriving from sharing of goods at a global level.

(2) Limited evaluation of the effects at an international level. This obstacle stems from the fact that every action taken by a single country has consequences for the ability of other countries to make progress on the various SDGs by considering some of their effects in the business context, such as the international development of finance and per capita emissions of GHGs.

(3) The inclusion of 'unofficial' indicators: this aspect is attributable to the fact that for many 'official' SDG indicators most of the data necessary for the calculation are missing; therefore, they have been replaced with unofficial metrics, even if drawn from reliable sources.

(4) Lack of time series data analysis: this gap is essentially due to the lack of historical data for most of the variables considered in the calculation of the various indicators.

The 2018 report incorporates several new indicators that were replaced or modified due to changes in methodological approach and estimates collected by data providers. For the first time, the 2018 report incorporates trend data to assess countries' progress towards meeting the 2030 SDG deadline. Using historic data, the 2018 report estimates:

how fast a country has been moving towards an SDG and determine whether – if continued into the future – this pace will be sufficient to achieve the SDGs by 2030. For each indicator, SDG achievement is defined by the green threshold set for the SDGs Dashboards. The difference in percentage points between the green threshold and the normalized country score denotes the gap that must be closed to meet that goal. (Bertelsmann Stiftung and Sustainable Development Solutions Network, 2018, p. 43).

For example, with respect to SDG 13 there are five indicators included in the 2018 dashboards reported in Table 1.

Table 1. List of Indicators for 2018 Index and Dashboards (SDG 13).

SDG	Indicator	Notes	United Nation Security Council list	Source	Description
13	Energy-related CO_2 emissions per capita (tCO_2/capita)		Not in UN-STATS database	Oak Ridge National Laboratory (2018)	Emissions of carbon dioxide per capita that arise from the consumption of energy. This includes emissions due to the consumption of petroleum, natural gas, coal and also from natural gas flaring
13	Imported CO_2 emissions, technology-adjusted tCO_2/capita		Not in UN-STATS database	Kander et al., 2015	Imports of CO_2 emissions measured as technology-adjusted, consumption-based (TCBA) emissions minus production-based emissions. Technology-adjusted emissions data reflect the carbon efficiency of exporting sectors. If a country uses relatively CO_2 intensive technologies in its export sector then it will have a higher TCBA than suggested by a simple carbon footprint
13	Climate Change Vulnerability Monitor (best 0–1 worst)		Not in UN-STATS database	HCSS (2015)	The index assesses global variations in vulnerability to climate change by gauging each country's vulnerability to three main potential impacts of global warming: increase in weather-related disasters, sea levels rise and loss of agricultural productivity

13	CO_2 emissions embodied in fossil fuel exports (kg/capita)	Not in UN-STATS database	UN Comtrade (2018)	Kilograms of CO_2 emissions per capita embodied in the exports of coal, gas and oil. Calculated using a three year average of fossil fuel exports and applying CO_2 conversion factors to those fossil fuels. When exports data for countries with little to no production of fossil fuels, we assumed a value of 0
13	Effective carbon rate from all non-road energy excluding emissions from biomass (€/tCO_2)	Not in UN-STATS database	OECD (2018)	Average effective carbon rates, the price of carbon emissions resulting from carbon and emissions trading systems excluding CO_2 emissions from biomass

Source: Bertelsmann Stiftung and Sustainable Development Solutions Network (2018, pp. 45–46).

NOTES

1. The authors would like to point out that this reference is not inspired by religious beliefs. The Encyclical Letter *Laudato Si* of the Holy Father Francis on care for our common home, has been analyzed highlighting the connections between corporate social responsibility of businesses, the importance of cooperative relationships to create value for companies and the personal responsibility' (Cremers, 2016) and reflects on the ethical dimension governing science, technology and risk regulation (Spina, 2015). http://www.vatican.va/content/francesco/en/encyclicals/documents/papa-francesco_20150524_enciclica-laudato-si.html.

2. For more details see International Energy Agency, 2013, 2015a, 2015b, 2016.

3

WHAT IS THE SDG 13?

3.1. SDG 13 AND THE MAIN FOCUS ON CLIMATE CHANGE

Every country in the world is affected by climate change. National economies are negatively influenced by climate change that is impacting on lives and creating increased costs for people, communities and countries. The consequences of climate change can be disruptive, involving changing weather patterns, rising sea levels and more life-threatening weather events (Church & White, 2006). One of the most important drivers of climate change can be identified in the GHG emissions from human activities. These emissions are continuing to increase and are now at their highest levels in history. Without action, the world's average surface temperature is projected to rise over the twenty-first century and is likely to surpass 3 °C this century – with some areas of the world expected to warm even more. The poorest and most vulnerable are being affected the most. As the United Nations explains:

> *Affordable, scalable solutions are now available to enable countries to leapfrog to cleaner, more resilient economies. The pace of change is*

quickening as more people are turning to renewable energy and a range of other measures that will reduce emissions and increase adaptation efforts. https://sdg-tracker.org/climate-change)

Climate change is a global challenge that does not respect national borders; emissions anywhere have an impact on people everywhere. It is an issue that requires solutions coordinated at the international level and international cooperation in helping developing countries move towards a low-carbon economy. Data and statistics on CO_2 and other GHG emissions can be found at Our World in Data (https://ourworldindata.org/co2-and-other-greenhouse-gas-emissions) with an overview of the increase of CO_2 emissions since the industrial revolution, disrupting the global carbon cycle and generating a planetary warming impact.

To address climate change, countries adopted the Paris Agreement at the COP21 in Paris on 12 December 2015. The Agreement entered into force shortly thereafter on 4 November 2016, and stated that all countries committed to limit global temperature rise to well below 2 °C, and given the high risks to strive for 1.5 °C.

SDG 13 focusses on climate change by defining five critical objectives that each country must recognise in presenting their specific responsibilities described in Table 2.

3.2. SDG 13 TARGETS AND INDICATORS

The United Nations has defined five targets and eight indicators. Targets describe the goals; indicators represent the metrics by which the world aims to track whether these targets are achieved. The original text of all the targets is given in the following, and the data on the agreed indicators are shown.

Table 2. Goal 13. Take Urgent Action to Combat Climate Change and its Impacts.

13.1	Strengthen resilience and adaptive capacity to climate-related hazards and natural disasters in all countries
13.2	Integrate climate change measures into national policies, strategies and planning
13.3	Improve education, awareness raising and human and institutional capacity on climate change mitigation, adaptation, impact reduction and early warning
13.4	Implement the commitment undertaken by developed-country parties to the United Nations Framework Convention on Climate Change to a goal of mobilising jointly $100 billion annually by 2020 from all sources to address the needs of developing countries in the context of meaningful mitigation actions and transparency on implementation and fully operationalise the Green Climate Fund through its capitalisation as soon as possible
13.5	Promote mechanisms for raising capacity for effective climate change-related planning and management in least developed countries and small island developing states, including focussing on women, youth and local and marginalised communities

Source: https://sdg-tracker.org/climate-change.

Target 13.1: Strengthen Resilience and Adaptive Capacity to Climate-related Disasters

UN definition: *Strengthen resilience and adaptive capacity to climate-related hazards and natural disasters in all countries.*

This target refers to natural disasters, on which more information can be found at Our World in Data (https://ourworldindata.org/natural-disasters). For example, data about the number of reported disaster events are given. They include those from drought, floods, extreme weather, extreme temperature, landslides, dry mass movements, wildfires, volcanic activity and earthquakes. It is possible to note a radical increase of these events from 2000 to date. Moreover, the annual number of deaths and the annual death rate are shown by some charts. Other human impacts from disasters, such as injury, homelessness and displacement, can have significant consequences for populations. There are also specific links to the following human impacts:

- Injuries: number of people injured is defined as 'people suffering from physical injuries, trauma or an illness requiring immediate medical assistance as a direct result of a disaster'.

- Homelessness: number of people homeless is defined as 'number of people whose house is destroyed or heavily damaged and therefore need shelter after an event'.

- Affected: number of people affected is defined as 'people requiring immediate assistance during a period of emergency, i.e. requiring basic survival needs such as food, water, shelter, sanitation and immediate medical assistance'.

- Total number affected: total number of people affected is defined as 'the sum of the injured, affected and left homeless after a disaster' (https://ourworldindata.org/natural-disasters).

It is worth noting that figures of the economic costs from natural disasters reached their highest value in 2010 at about $350 billion.

SDG Indicator 13.1.1 is on 'Deaths and injuries from natural disasters' and it has been defined as 'the number of deaths, missing persons and directly affected persons attributed to disasters per 100,000 population'. Indicators measured here report mortality rates, internally displaced persons, missing persons and total numbers affected by natural disasters.

SDG Indicator 13.1.2 is focussed on 'National disaster risk management'. Indicator 13.1.2 is the number of countries that adopt and implement national disaster risk reduction strategies in line with the Sendai Framework for Disaster Risk Reduction 2015–2030. This indicator identifies countries which have/have not adopted and implemented disaster risk management strategies in line with the Sendai Framework for Disaster Risk Reduction.

SDG Indicator 13.1.3 is on 'Local disaster risk management'. This indicator has measured the proportion of local governments that adopt and implement local disaster risk reduction strategies in line with national disaster risk reduction strategies.

Target 13.2: Integrate Climate Change Measures into Policy and Planning

UN definition: *Integrate climate change measures into national policies, strategies and planning.*

SDG Indicator 13.2.1 is on 'Integration of climate change into national policies'. This indicator has been defined as the number of countries that have communicated the establishment or operationalisation of an integrated policy/strategy/plan which increases their ability to adapt to the adverse impacts of climate change and foster climate resilience and low GHG emissions development. This indicator measures the number of countries signed up to multilateral agreements on climate change. Currently, it does not reflect the levels of operationalisation or implementation of climate mitigation and adaption action. National commitments within the UNFCCC Paris Agreement vary by country depending on their NCDs hence they are not directly comparable. In the additional charts on the website it is possible to find data on national CO_2 emissions, per capita emissions and carbon intensity measures.

Target 13.3: Build Knowledge and Capacity to Meet Climate Change

UN definition: *Improve education, awareness raising and human and institutional capacity on climate change mitigation, adaptation, impact reduction and early warning.*

There are two indicators: the first is *SDG Indicator 13.3.1* which is focussed on 'Education on climate change', the definition being the number of countries that have integrated mitigation, adaptation, impact reduction and early warning into primary, secondary and tertiary curricula; the second, *SDG Indicator 13.3.2*, relates to 'Capacity-building for climate change'. This indicator is the number of countries that have communicated the strengthening of institutional, systemic and individual capacity-building to implement adaptation, mitigation and technology transfer, and development actions.

Target 13.A: Implement the United Nations Framework Convention on Climate Change

UN definition: *Implement the commitment undertaken by developed-country parties to the United Nations Framework Convention on Climate Change to a goal of mobilising jointly $100 billion annually by 2020 from all sources to address the needs of developing countries in the context of meaningful mitigation actions and transparency on implementation and fully operationalise the Green Climate Fund through its capitalisation as soon as possible.*

SDG Indicator 13.A.1 is about the 'Green Climate Fund mobilization of $100 billion'. The definition of this indicator is the mobilised amount of US dollars per year between 2020 and 2025 accountable towards the $100 billion commitment. This indicator measures the current pledged commitments from countries to the Green Climate Fund (GCF) as annual US$ contributions pledged. Also shown is the collective global total. Unlike most SDG targets which have set a target year of 2030, this indicator requires a mobilisation of $100 billion per year from 2020 onwards. There is also an additional chart on GCF pledges per capita, 2018.

Target 13.B: Promote Mechanisms to Raise Capacity for Planning and Management

UN definition: *Promote mechanisms for raising capacity for effective climate change-related planning and management in least developed countries and small island developing states, including focussing on women, youth and local and marginalised communities.*

SDG Indicator 13.B.1 refers to 'Support for planning and management in least-developed countries'.

Indicator 13.B.1 has been defined as the number of least-developed countries and SIDS states that are receiving specialised support, and the amount of support, including finance, technology and capacity-building, for mechanisms to raise the capacity for effective climate change-related planning and management.

SDG 13 requires urgent action on climate change that is central to delivering sustainable development (UNDP, 2016). Tackling climate change through adaptation and mitigation can represent a great tool to drive sustainable development outcomes in the key areas of Agenda 2030. In many cases, action on climate change can generate additional positive effects in terms of sustainable development. For example, mitigation and investments in renewable energy can produce development impacts in terms of an improvement of energy access, health benefits due to the reduction of polluting emissions and the creation of new jobs determined by investments in the renewable energy sector. Action to protect forests and land use brings advantages for the climate change and, at the same time, positively affects sustainable livelihoods for communities. By the same token, enhancing and building resilient economies and action to reduce poverty can have climate impacts through adaptation measures. Moreover, other positive impacts in terms of sustainable development are the reduction of disaster risk, advancing gender equality and providing basic social services. Measures on climate change can also improve social issues such as fragility, displacement, migration and conflict.

It is worth noting that climate change is 'a cross-cutting issue' to be addressed in order to achieve a successful implementation of all 17 SDGs. For example, Goal 1 (ending poverty) and Goal 11 (sustainable communities and cities) have their own specific targets, but other SDGs such as SDGs 7, 12, 14 and 15 support national action to combat climate change and to enhance zero-carbon growth (UNDP, 2016). Agenda 2030 has identified a focussed goal, that is, SDG 13, to promote climate-specific actions in adaptation, mitigation and finance at a global level. Issuing a stand-alone goal on climate change has been essential, but it has to be considered in connection with the other goals. To ensure major implementation of goal 13 by governments in different countries, it is important to implement climate actions through other goals by sustainable development approaches.

To support countries in reaching the SDG 13 targets, it is important to adopt climate measures in three key action areas, that is, *climate change adaptation, zero-carbon development* and *scaled-up climate finance*. The first action area aims to put in place policies and measures to decrease vulnerability and risks linked to the impacts on sustainable development. Examples include support for national adaptation planning to identify and address potential impacts, community-based measures to build resilient livelihoods and improve food and water security for agricultural communities affected by changing rainfall patterns, climate-informed social protection systems and disaster risk governance and access to climate information and early warning systems to ensure development is risk-informed and addresses the growing frequency and intensity of natural hazards (UNDP, 2016, p. 5).

The second action area provides initiatives to reduce the existing level of carbon missions suggesting different policies. Zero-carbon development examples include investment in renewable energy solutions such as solar or wind power, and pursuing energy efficiency measures that reduce emissions from energy intensive activities. Other measures aim to combat deforestation and enhance protection of forests, wider land use issues including agriculture and land degradation, as well as other emission sources, including methane. Policies on Goal 7 on energy and Goal 15 on ecosystems, forests and biodiversity will play an important complementary role in advancing Goal 13 on climate change action (UNDP, 2016, p. 5). The final action area is focussed on scaled-up finance in order to ensure that sufficient resources are available for emerging economies to tackle climate change. Moreover, it is essential to induce countries to increase the amount of climate actions through the adequate budgeting, planning and spending of national and international funds.

Climate change does not affect every country in the same way; there may be different responses from different areas depending on the geographical location, vulnerability and the capacities of countries. Some factors depend on national and local governments. It is important to highlight the role of the contextualisation of climate actions. The level of action and type of interventions must be nationally contextualised and adequate to the climate context and risk profile of a country or areas within a country (UNDP, 2016, p. 6).

Given its extensive experience, UNDP can help countries with a set of policies and programme support to tackle climate change and go beyond the aspirations of national climate pledges. A summary of these initiatives is shown in the report (from page 9), highlighting specific actions for each SDG 13 target. For example, there is a summary of climate change impacts ('key facts') highlighting causes, global impacts, regional impacts and action (UNDP, 2016, p. 7).

4

CLIMATE CHANGE TRENDS: UNDERSTANDING SDG 13 LOCAL AGENDA AND KEY INTERACTIONS FRAMEWORKS

4.1. CLIMATE CHANGE AGREEMENT SIGNED IN DECEMBER 2015 IN PARIS (COP21)

In December 2015, the agreement on climate change (COP21) was signed in Paris by 196 states inviting all countries to accelerate and intensify the actions and investments necessary for a sustainable future, low-carbon economy resilient to climate change.[1] The agreement on climate change reached in Paris has set itself the ambitious objective of maintaining [...] the increase in the average global temperature well below 2 °C compared to pre-industrial levels and to continue efforts to limit this increase to 1.5 °C compared to pre-industrial levels, and is certainly an unprecedented agreement in an attempt to limit global warming by reducing GHG emissions, that include 'direct emissions' that come from a company's own sources or those controlled by a company (Scope 1) and 'indirect emissions' that are a consequence of a company's activities but whose source is controlled by other companies (Scopes 2 and 3).

- Carbon dioxide (CO_2) 82%: Carbon dioxide enters the atmosphere through the combustion of fossil fuels (coal, natural gas and oil), solid waste, trees and wood products and is the result of certain chemical reactions (e.g., cement manufacture). Carbon dioxide is removed from the atmosphere when it is absorbed by plants through the biological carbon cycle.

- Methane (CH_4) 10%: Methane comes from the production and transportation of coal, natural gas and oil. Methane emissions also derive from livestock, other agricultural practices and the decay of organic waste in urban solid waste landfills.

- Nitric oxide (N_2O) 5%: Nitric oxide is produced during agricultural and industrial activities, as well as during the combustion of fossil fuels and solid waste.

- Fluorinated gases 3%: Hydrofluorocarbons, perfluorocarbons, sulphur hexafluoride and nitrogen trifluoride are synthetic and powerful gases emitted by a series of industrial processes. These gases are typically emitted in modest amounts, but since they are powerful GHGs, they are sometimes referred to as high-power gas global warming ('gas high GWP').

4.2. INTERACTIONS FRAMEWORKS

Pradhan et al. (2017) established pairs of SDGs and correlated their interaction in different countries, mapping and counting those populations more predisposed to act in these pairs mainly because they feel that some are currently more urgent than others. They classify pairs as synergy pairs where actions in one objective favours another; trade-off pairs are defined as those where interaction in one objective will jeopardise the progress of another.

The pair SDG 13 and SDG 11 comes in at first place in the top 10 synergies, while not figuring in any pair of trade-offs, which implies its importance. The sharing of indicators also contributes to the synergy revealed between SDGs 13 and 11, in the opinion of the authors.

Regarding the interactions of SDGs between countries, this does not appear in any pair in the top 10, begging the question as to whether it is recognised as important for the process of change, or the recognition that there is still a long term to go through in SDG 13, the most cited and present in more pairs. In any case, it requires attention because this lack of engagement from the citizenship with SDG 13 may compromise the 2030 agenda's success.

An interesting tool for evaluating SDGs interaction is a document issued by the International Council for Science (2017, *A Guide to SDG Interactions – From Science to Implementation*) which investigates the nature of links between the SDGs. This report is based on the premise that a science-informed analysis of interactions across SDG domains – which is currently lacking – can support more coherent and effective decision making, and better facilitate follow-up and monitoring of progress.

The achievement of long-lasting, sustainable development outcomes can be facilitated by understanding possible trade-offs as well as synergistic relations between the different SDGs. The guide suggests a scoring approach to stimulate more science-policy dialogue on the importance of interactions to provide a starting point for policymakers and other stakeholders. The aim is to set priorities and implementation strategies, and to engage the policy community in further knowledge developments in this field. The document analysed four SDGs, 2, 3, 7 and 14, and found that there are synergies with the other SDGs. For example, the interaction between SDG 2 and SDG 13 highlights the following considerations:

> *Agriculture is an important source of greenhouse gas emissions and so contributes to climate change. Conversely, climate change has wide-ranging impacts on agriculture and food security through extreme weather events as well as long-term climatic changes (such as warming and precipitation changes) and will significantly constrain the achievement of SDG 2. Sustainable agricultural practices play an important role in climate adaptation and mitigation (such as improving soils and land quality, genetic diversity, and bioenergy). (International Council for Science, 2017, p. 11)*

In summary, there is a 'bidirectional' link between climate change/SDG 13, as the threat to food production can be seen as the greatest driver of climate change, yet sustainable agriculture can be part of the solution. Transformative action in food systems is needed to achieve SDG 13 (and other SDGs), involving technical, policy, capacity enhancement and finance elements. At the same time transformative actions can determine risks for farmers, investors, development agencies and politicians (Campbell et al., 2018).

Other important interactions are between SDG 3 and SDG 13 on the impact of climate change on health. Many of these impacts are direct (such as the effects of heat stress on the ability to work outside), while others are indirect, arising from climate change enabling the spread of disease or contributing to food and water insecurity or to mass movements of people. The health goal cannot be achieved if there is a failure to address the climate action goal. Interaction with

SDG 7 relates to the increase in renewables and improvement in energy efficiency as part of efforts to keep global warming well below 2 °C above pre-industrial levels. In this regard, it is possible to consider SDG 7 as 'enabling factor' for the implementation of SDG 13. Some studies have pointed out a high degree of interaction (Nilsson, Griggs, & Visbeck, 2016; Nilsson et al., 2018) especially between 7.2 (renewables) and 7.3 (efficiency) with 13.1 adaptation. For example, technology innovation investments towards SDG 7 can relate to the digitation of electricity grids that could decrease fuel consumption with environmental benefits for SDG 13. Finally, interaction with SDG 14 shows the mutual influence of climate change on oceans and coastal ecosystems. Thus, achieving SDG 14 and SDG 13 is highly synergistic, for example, through the conservation of coastal ecosystems acting as blue carbon sinks. It is important to manage carefully the need to ensure that climate adaptation and coastal and marine protection measures do not conflict.

NOTE

1. http://www.accordodiparigi.it/

5

PRACTICAL TOOLS AND MECHANISMS FOR SDG 13 IMPLEMENTATION

5.1. SETTING THE LOCAL AGENDA: GOVERNMENTS AND PUBLIC SECTOR

There is no country in the world that is not experiencing first-hand the drastic effects of climate change. GHG emissions continue to rise; they are now more than 50% higher than their 1990 level. Furthermore, global warming is generating long-lasting changes to the climate system which threaten irreversible consequences if countries and governments do not take action now. Given these considerations, the crucial question is: are governments organising for the SDGs?

Between February and May 2018, SDSN and the Bertelsmann Stiftung (Bertelsmann Stiftung and Sustainable Development Solutions Network, 2018) carried out a short survey to gauge political leadership and the institutionalisation of the SDGs in G20 countries. The survey was articulated in 15 questions and a number of sub-questions. It covered the following aspects: (1) national strategy and baseline assessments in the executive, (2) coordinating units in the executive and (3) budgeting practices in the executive. Additional questions were also included on legislative actions, as well as the main challenges for implementation.

Evidence from this survey demonstrates that there are significant variations among the G20 countries regarding the institutionalisation of the SDGs. For example, some countries such as Brazil, Italy and Mexico show relatively high levels of institutionalisation. This is characterised, for instance, by some factors such as the existence of SDG strategies and action plans, coordination units in government tasked with supporting the implementation of the goals and stakeholder engagement tools such as SDG web platforms and portals. On the contrary, countries such as the United States and the Russian Federation demonstrated low levels of political leadership and institutionalisation of the SDGs. This aspect is characterised largely by the absence of public statements made by the head of state on how the country plans to implement the SDGs (Bertelsmann Stiftung and Sustainable Development Solutions Network, 2018, p. 2).

Given the importance of the long-term pathway to achieve the SDGs, a crucial factor is the development of shared action by mobilising stakeholders around shared strategies and identifying innovation challenges.

An interesting analysis has been carried out by the Climate Action Tracker (CAT) and Climate Transparency on both the content of intended nationally determined contribution (*what governments propose to do*) and current policies (*what governments are actually doing*) on climate mitigation in G20 countries (Bertelsmann Stiftung and Sustainable Development Solutions Network, 2018, p. 2). These independent reviews of SDG 13 carried out by experts show that, with the exception of India, NDCs and current climate policies tracked by G20 countries are scarce and, in some cases, critically insufficient to achieve the 'well below 2 °C' objective of the Paris Climate Agreement. Some countries have issued insufficient targets which they can reach without implementing new policies. Others have implemented policies that will not even allow insufficient targets to be met. These findings are developed from independent reviews applied to climate action that focussed on both the content of strategies and the alignment of policy actions. This analysis can provide an interesting insight into the need to move beyond published action plans and mere intentions. It is possible to apply the methodology used by the CAT in countries where the SDGs have already been translated into more tangible short- and long-term national targets. This method can offer the opportunity to evaluate the level of ambition and policy actions targeting other SDG priorities, such as reducing income inequality, universal health care, sustainable land-use and food systems or international development finance. From this perspective, it is becoming crucial to strengthen the accountability system in order to measure progress on sustainable development to complement gross domestic product.

This survey demonstrated that a majority of G20 countries have developed or are in the process of developing national indicators of progress on the SDGs. It is worth noting that there is no common approach for identifying the nature and number of national indicators to control progress on the SDGs, which range from 63 in Germany to 201 in Italy. The European Union (EU), via Eurostat, has identified 100 indicators to monitor the implementation of the SDGs in the EU.

Regular audits can also support accountability and government. Some supreme audit institutions (SAIs) have started to undertake performance audits in the context of the SDGs. Traditionally SAIs have focussed on financial and compliance audits, but they are increasingly incorporating other aspects, such as performance and value for money audits (OECD, 2015).

The International Organisation of Supreme Audit Institutions has issued guidelines for how SAIs can contribute to the success of the 2030 Agenda and the SDGs in their countries.

Another important factor is represented by some public management tools such as budgeting practices and procedures. Performance budgeting and spending reviews are powerful tools that can support mainstreaming the SDGs in the budget process and strengthen implementation over time at the national level. The achievement of long-term goals can also be facilitated by three examples of other government tools and functions: (1) incorporating sustainable development principles into regulatory governance, (2) the strategic use of public procurement and (3) mobilising the machinery of government for the SDGs from the adaptation of Human Resource Management practices.

Given the lack of comparative country-level information regarding the integration of sustainable development in public management processes, further analysis is needed to identify best practices and to adopt innovative approaches.

5.2. LOCALISATION OF THE SDGs

Localisation means considering sub-national contexts, challenges, opportunities and governments at all stages of the Agenda, from the setting of goals and targets, to determining the means of implementation and using indicators to monitor progress. Local governments were not involved in the negotiation of the MDGs, even though many of the areas covered by the goals are local government's responsibilities. The lack of local ownership of the goals, as well as insufficient resources at a local level to implement them, was identified as major weaknesses of the MDGs. Now

Agenda 2030 needs to be localised. For United Cities and Local Governments (UCLG), the Global Taskforce of Local and Regional Governments, the 'localisation' process started with UCLG President Topbaş' participation in the Secretary-General's High-level Panel of eminent persons on the Post-2015 Development Agenda. The inclusion of Mayor Topbaş in the High-level Panel gave sub-national governments a voice to advocate for the importance of local governance and local resources at the highest level UCLG and other partners, successfully campaigned for a stand-alone goal on sustainable cities and human settlements (SDG 11). UCLG also proposed localised targets and indicators and advocated for targets and indicators that would differentiate between urban and rural contexts (see the technical report: How to localise targets and indicators of the Post-2015 Agenda, https://www.uclg.org/en/issues/2030-agenda-sustainable-development).

ᐧRegarding climate change, UCLG has been constantly involved in climate change negotiations, and the organisation has raised awareness on the impact of climate change on cities and their inhabitants, and promoted the implementation of a sustainability agenda to prevent its impact around the world.

Recent milestones (from 2015 to date) are reported as follows.

2018 – Local elected leaders meet at COP24 in Katowice, Poland to continue climate negotiations in the spirit of the Paris Agreement. Scientists, policymakers, urban practitioners and local and regional government networks gather at the IPCC Cities and Climate Change Science Conference in Edmonton, Canada, from 5th March to 7th March to inspire the next frontier of research focussed on the science of cities and climate change.

2017 – UCLG, as the largest network of local and regional governments, continues to support the Global Covenant of Mayors for Climate and Energy launched in January 2017.

This coalition, resulting from the merger of the Compact of Mayors and European Covenant of Mayors, is the largest initiative for the reduction of local GHG emissions, which aims to foster resilience to climate change and track its progress. Following recommendations by the Policy Council on Resilience, UCLG members gather in Hangzhou at the World Council, welcome the initiative presented by Ronan Dantec to celebrate the One Planet Summit on the occasion of the second anniversary of the Paris Agreement, and to highlight the need to enhance commitments from all actors, as well to pay particular attention to the financing of initiatives to fight climate change.

2016 – With the entry into force of the Paris Climate Agreement, UCLG actively contributes to the Climate Summit for Local and Regional Leaders, held in the presence of the major regional and international networks of cities and local and regional authorities in the framework of COP22 in Morocco. Its outcome document, the Marrakech Roadmap for Action, calls for a global action framework towards localising climate finance.

2015 – Thousands of mayors and local leaders come together in Paris to strengthen the voices of local and regional governments at the COP21. Local and regional leaders emphasise the need to link climate action to the 2030 and Habitat III Agendas, and reaffirmed the importance of citizen participation and inclusive social policies to ensure the effective implementation of the sustainability agenda in its broadest sense. Through the Paris Agreement, the UNFCCC Parties have recognised that 'adaptation is a global challenge faced by all with local, subnational, national, regional and international dimensions'.

In particular, climate actions related to SDG 13 are being carried out by local and regional governments by driving for change in global negotiations for over two decades (UCLG, 2019).

For example, at the Global Climate Action Summit in September 2018, a number of local and regional governments committed to new, more ambitious targets for zero-emission transport, the use of 100% renewable energy, net zero-carbon buildings and zero-waste by 2030. The transition to a zero emissions economy is an urgent need but it requires political leadership at all governmental levels. It is worth noting that the links between climate change (SDG 13), consumption and production (SDG 12) and inequalities (SDG 10) are particularly observable at urban and regional levels. In this regard, many local and regional leaders are attempting to adopt both climate-related and social goals driving some policies and inclusive actions. For example, in Wales there is a requirement for public bodies, called in Welsh the 'Well-being of Future Generations Act', to include long-term planning to meet goals such as resilience. A similar policy is adopted in Spain by the Basque government that includes climate change as a major topic, in its territorial plan to raise the quality of life of citizens, while the autonomous government of Catalonia passed a law in 2017 that included a provision to keep vulnerable populations from energy and water poverty. The main aim is to strengthen the urban-climate nexus by embedding policy coherence in sustainable low-carbon urban and territorial development and implementing vertically aligned NDC investment plans (ULGC, 2019). In summary, there are several claims[1] to national governments and to the UN system to advance the implementation of SDG 13 at all levels of government. In particular, a crucial role in combating climate change is played by cities. Cities have a unique ability to address global climate change challenges. The extent and impact of climate change, as well as the ability to achieve emission reductions and the capacity to adapt to changing circumstances, are strongly influenced by the choices made on well-established urban infrastructure. Policy frameworks should be well-aligned and work to support city-level action on climate change (OECD, 2014). The availability of funding and finance for low-carbon projects in cities represents a crucial issue for the Paris Agreement and 2030 Agenda. One of the most important organisations, C40 Cities, which connects the world's greatest cities to deliver climate change, pointed out some important achievements in its latest report (McKinsey Centre for Business and Environment, C40 Cities, 2017). By 2020, the most crucial commitment for all C40 members is to have a comprehensive, measurable climate action plan in place to deliver low-carbon resilient development consistent with the 1.5 °C target of the Paris Agreement. A total of 25 pioneering cities signed this commitment in 2017 and C40 prepared a document *Focused acceleration – A strategic approach to climate action in cities to 2030* in partnership with The McKinsey Centre for Business and Environment, which provides a delivery roadmap with guidance for cities on how to reduce emissions in line with the ambitions of the Paris Agreement. A total of 53 C40 cities have inventories that are compliant with the Global Protocol for Community-scale Greenhouse Gas Emission Inventories (GPC) standard. To support cities in developing GPC compliant inventories, C40 launched the City Inventory Reporting and Information System tool – an accessible, easy-to-use and flexible Excel-based tool for managing and reporting city GHG inventory data. Moreover, C40 developed the Climate action for URBan sustainability (CURB) scenario planning tool in order to provide 'strategic-level' analysis to help cities identify and prioritise low-carbon infrastructure and other GHG emissions reduction actions (C40, 2017). In particular, to overcome financing barriers, several city-focussed project preparation funds have established some pipelines of bankable climate projects such as the Cities Development Initiative for Asia, C40 Cities Finance Facility and ICLEI's Transformative Actions Programme. These initiatives can stimulate networking opportunities by establishing connections with important financiers. In summary, policy recommendations to national governments and cities aim to promote a resource-efficient, circular and waste-free society by fostering greater civic participation and more inclusive societies.

5.3. INCLUSIVENESS AND THE PARTICIPATORY APPROACH

National and sub-national governments, civil society organisations, the private sector, academia and individual citizens should all be involved in the implementation and monitoring of the SDGs. Sub-national governments in particular can play a crucial role in raising awareness about Agenda 2030 and its relevance for local communities. In this way they bridge the gap between central governments and communities and play a strong role in fostering the involvement of civil society organisations, the private sector (micro, small and medium enterprises), academia and other community-based organisations.

One of the most important mechanisms for ensuring the successful implementation of Agenda 2030 at local level is the adoption of democratic accountability through a participatory approach. Awareness-raising activities should aim to increase the engagement of citizens and local communities in order to promote their sense of ownership of the Agenda and their involvement in the achievement of the SDGs at a local level. To stimulate awareness raising, it is not enough to inform citizens about the existence of the SDGs; it is also important to act on empowering them to participate in the achievement of the SDGs in their daily lives. With this approach, municipal and regional governments should be supported in recognising the 2030 Agenda as a framework for action, and set up mechanisms that enable citizen participation and institutional accountability.

An open participatory approach can represent an important tool to better implement sustainable development principles and practices; stakeholders' involvement (experts and non-experts) can facilitate the process of gathering data, and the design and implementation of a collaborative report (Ramos et al., 2014).

5.4. PRIVATE SECTOR AND BUSINESS CONTEXT

COP21 and the SDGs represent an important step in the transition towards better and more exhaustive non-financial corporate reporting that private organisations and companies use to communicate – to investors, credit and sustainability rating agencies, insurance companies, credit institutions and other stakeholders – further standardised information on their exposure to climatic risks. It is, therefore, necessary to integrate the existing corporate reporting models to include non-financial information on environmental and social impacts, and to connect financial capital with natural capital (Climate Disclosure Standard Board, 2015). To do this, it is possible to use a framework that also allows communication of 'investment grade' environmental and social data, namely complete, consistent, reliable, comparable and transparent, with the same consistency as financial ones. In turn, this framework helps provide investors with useful and usable information for the decision-making process on sustainability, thus improving value creation. Companies must communicate clear and concise non-financial information to investors, who increasingly also consider climate risks as a critical factor with an impact on corporate value and their decisions (Gasperini & Doni, 2017a, 2017b, 2017d). Investment decisions with potentially serious consequences for the environment, particularly over water consumption, waste management and GHG emissions, can cause value destruction in the medium and long term.

5.5. PLANNING FOR SDG 13 IMPLEMENTATION: CLIMATE RISK ASSESSMENT

Environmental, social and governance (ESG) factors and their impact on sustainability are gaining greater attention from financial markets, are less niche and more mainstream for asset managers and financial analysts (Gasperini, 2013; Nagy et al., 2015). That said, there is a growing interest in sustainable finance by institutional investors who have a long-term vision, such as pension funds, sovereign wealth funds, insurance companies and, more recently, also by religious orders. Long-term investors, who have always paid attention to value creation, now also require information on climate risks and natural capital, as these issues affect their investment decisions. This behaviour indicates a move away from initial scepticism about the link between socially responsible investments (SRI) and unfavourable financial performance (see, e.g., the document issued by European Fund and Asset Management Association, EFAMA, 2016, or the study by Friede, Busch, & Bassen, 2015, from which the authors argued that 'there is no statistically relevant outperformance or underperformance of Responsible Investment strategies').

Many investors are focussing their attention on companies working to mitigate the current effects of climate change by adopting disinvestment strategies from fossil fuels. These companies are aware of the risks and opportunities that a transition to a low-carbon economy can cause; its production processes are not the cause of high emissions of GHG, therefore investments in financial instruments such as low-carbon

Exchange-traded Funds (ETF) and environmental thematic funds are beginning to attract interest.

Since the 1950s, many of the impacts on human and natural systems generated by global warming have been unprecedented, as several studies confirmed. The main consequences for human and natural systems generated by the climatic warming are the following:

- *Sea level*: a sea level rise of about 17 cm (6.7 in.) in the last century and an increase in global temperature whose rate in the last decade is almost twice that of the previous century (Church & White, 2006).

- *Global temperature*: from various temperature reconstructions it appears that the global surface of the earth has undergone an increase since 1880; the greatest increase has occurred since 1970, with the 20 warmer years that have been found since 1981 and the next 10 years have been the hottest in the last 12 years (Peterson et al., 2009).

- *Ocean warming*: the oceans absorbed much of this increase in heat, with the upper 700 m (about 2,300 feet) showing a warming of 0.302 F since 1969 (Levitus et al., 2009).

- *Ice caps*: the mass of Greenland and the Antarctic ice sheets has decreased. According to NASA's Gravity Recovery and Climate Experiment research data, Greenland has lost between 150 and 250 km^3 (36–60 cubic miles) of ice in the years between 2002 and 2006, while Antarctica has lost about 152 km^3 (36 cubic miles) of ice between 2002 and 2005.

- *Arctic sea ice*: both the extent and thickness of Arctic sea ice has diminished significantly over the past few decades (Polyak et al., 2009, chapter 7).

- *Glacier retreat*: Glaciers are retreating almost everywhere in the world – including in the Alps, the Himalayas, the Andes, the Rocky Mountains, Alaska and Africa (the National Snow and Ice Data Centre and World Glacier Monitoring Service).

- *Extreme events*: the number of high temperature events in the United States is growing, while the number of low temperature events has been decreasing since 1950. In the United States, there has also been an increase in the intensity of rainy events (http://lwf.ncdc.noaa.gov/extremes/cei.html).

- *Ocean acidification*: From the beginning of the industrial revolution, the acidity of the surface of the ocean waters has increased by about 30%. This increase is the result of activities that emit more carbon dioxide into the atmosphere and therefore a larger share is absorbed by the oceans. The amount of carbon dioxide absorbed by the upper layer of the oceans increases by about 2 billion tons per year (Sabine et al., 2004).

- *Reduction of snow cover*: satellite observations have revealed that the amount of snow cover in the northern hemisphere has decreased over the past five decades due to snow melting (Derksen & Brown, 2012; National Snow and Ice Data Centre, Rutgers University Global Snow Lab, Data History, http://nsidc.org/cryosphere/sotc/snow_extent.html, Accessed 29 August 2011).

This aspect is shown in Fig. 1 indicating CO_2 levels during the last three glacial cycles, reconstructed by an analysis of the surface of the ice carried out by NASA during the project *Global climate changes – Vital signs of the planet* (https://climate.nasa.gov/vitalsigns/carbon-dioxide/).

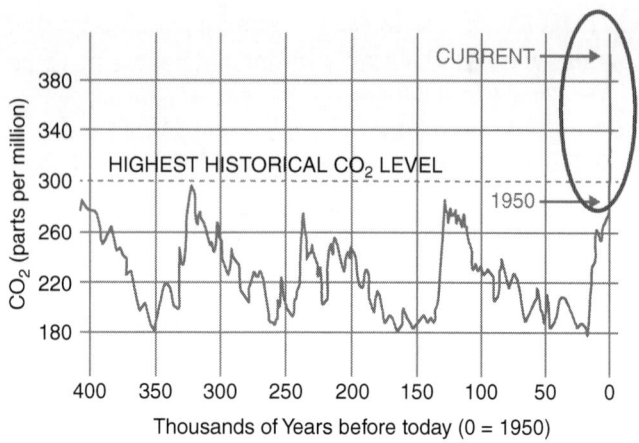

Fig. 1. Proxy (Indirect) CO_2 Measurement.

Source: Vostok ice core data/ Petit et al., 1999; NOAA Manua Loa CO_2 record.

The report published by IPCC Institute (2014), a leading international organisation for the assessment of climate change, highlights that human influence on the climate is clear, and the recent anthropogenic emissions of GHG are the highest in history. IPCC also emphasises scientific evidence that unequivocal global warming is taking place because of many observed events that have a widespread impact on human and natural systems.

A continuous increase in GHG emissions will be the cause of further warming and lasting changes in all components of the climate system, increasing the likelihood of acute, pervasive and irreversible risks for people and ecosystems. Limiting climate change is possible through a substantial reduction in GHG emissions that, together with adaptation, leads to a reduction in the impacts on human and natural systems.

Anthropogenic GHG emissions are mainly determined by population size, economic activities, lifestyles, energy use, land use, technology and climate policy. The IPCC Institute has therefore indicated several representative concentration paths (RCPs) of GHGs used to make projections based on these factors that describe four different trajectories in the twenty-first century for atmospheric concentration, pollutant gas emissions and the use of soil (Fig. 2).

The different RCP pathways were defined using an integrated assessment model as an input into a wide range of simulations used to project the consequences on the climate system. These climate projections, in turn, make it possible to define impacts and evaluate adaptation.

Approximately 300 basic scenarios and 900 mitigation scenarios classified by CO_2 equivalent (CO_2-eq) have been defined by 2100, including a strict mitigation scenario (RCP2.6), two intermediate scenarios (RCP4.5 and RCP6.0) and a scenario with very high GHG emissions (RCP8.5). Scenarios without further efforts to limit emissions ('basic scenarios') lead to pathways ranging from RCP6.0 to RCP8.5.

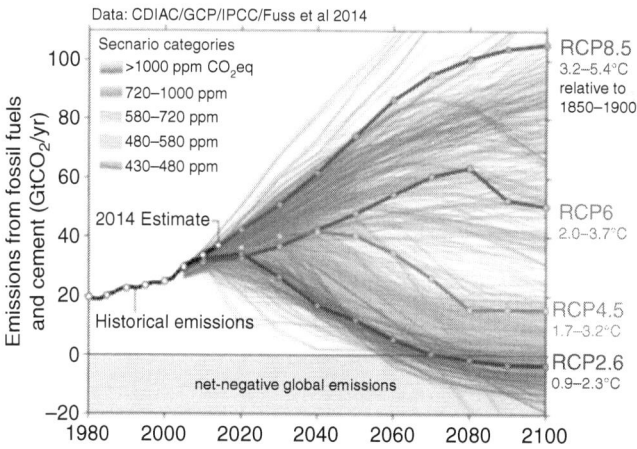

Fig. 2. Total GHG Emissions in IPCC Scenarios.

Source: Global Carbon Budget 2014 (Global Carbon Project).

RCP2.6 is the representative of a scenario that aims to maintain global warming below 2 °C compared to pre-industrial temperatures. Most models indicate that meeting scenarios forcing RCP2.6 levels are characterised by net negative emissions by 2100, on average around 2 $GtCO_2$/yr.

Risks due to an economic transition to a low-carbon economy imply opportunities, but also political, legal, technological and market risks that need to be addressed to monitor the requirements of mitigation and adaptation imposed by climate change. Climate change is now recognised as a major financial threat. A growing number of investors are aware that climate change may have an impact not only on the planet but also on their financial performance, if they are not able to evaluate correctly the risks. Depending on the nature and rapidity of climate change, the current economic transition involves organisations' understanding level of the following transition risks described in the *Final Report – Recommendations of the Task Force on Climate-related Financial Disclosures* (June 2017) that the financial markets must also be able to understand (Fig. 3).

- *Political risks*: policy decisions on climate change are constantly evolving and can be distinguished between decisions limiting actions that contribute to the adverse effects of climate change, and decisions necessary to promote adaptation to climate change.

- *Legal risks*: the failure of initiatives taken by organisations to mitigate the impacts of climate change, non-adaptation to climate change and insufficient information on material financial risks may lead to an escalation to legal disputes.

- *Market risk*: the transition to a low-carbon economy is creating important opportunities in all economic sectors where there is a growing demand for 'green products

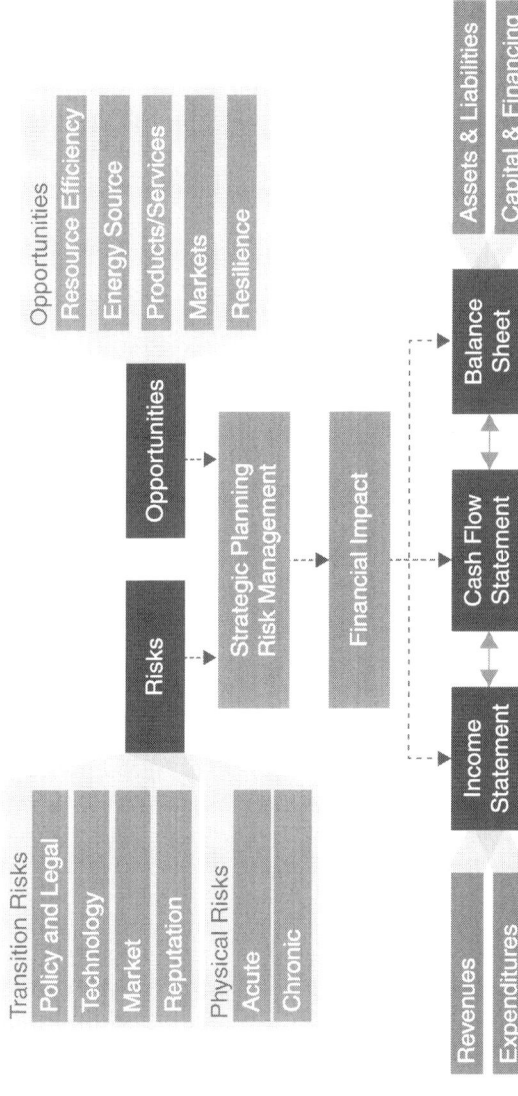

Fig. 3. Climate-related Risks, Opportunities and Financial Impact.

Source: Final Report – Recommendations of the Task Force on Climate-related Financial Disclosures (2017).

and services'; financial organisations not investing in innovation to take advantage of these opportunities are subject to competitive disadvantages.

- *Technological risk*: technological innovations in support of the transition to a low-carbon economy have a significant impact on organisations. New sources of renewable energy, battery accumulation, energy efficiency and carbon storage, etc. affect the competitiveness of organisations and their production and distribution costs. The timing of development and sharing of technology imply fundamental uncertainty in the assessment of technological risk.

- *Reputational risk*: given the financing of carbon intensive projects by the banking sector and the production of coal-based energy, and more public awareness and increasing concern about climate change, it is reasonable to assume that reputational risks will also increase.

To better identify the information required by investors, asset managers, financial analysts, lenders and insurance underwriters, and to adequately assess climate risks and opportunities, the Financial Stability Board (FSB) (http://www.fsb.org/) has set up a working group called Task Force on Climate-related Financial Disclosures (TCFD, 2016, 2017) whose task is to promote voluntary and coherent financial communication about climate change. This type of disclosure allows an understanding of material risks and supports to the stakeholders to better assess the concentration of carbon-related assets in the financial sector and the exposure of the financial system to climate risks.

The FSB Task Force published its *Final Report – Recommendations of the Task Force on Climate-related Financial Disclosures* (29 June 2017), which focuses on the analysis of the financial impacts of risks and climate opportunities on organisations rather than on the impact of the activities of an organisation on the environment. This report has received support from over 100 companies operating worldwide in many sectors.

The main elements of the recommendations are indicated as follows:

(a) *Positioning of the communication:*

- TCFD recommends that organisations provide climate-related financial information in their annual public financial documents.

(b) *Principle of materiality:*

- The recommendations concerning information on strategy, metrics and targets must be subject to a materiality assessment.

- The recommendations concerning information on governance and risk management activities must be disclosed as many investors want to understand the context of governance and risk management in which the financial and operational results of organisations are achieved.

(c) *Scenario analysis:*

- The TCFD encourages the communication of long-term information through a scenario analysis, as it is considered a useful tool for improving the resilience and flexibility of strategic plans.

- Many investors want to understand how organisations' strategies are resilient to climate risks.

- The recommendation for communications on strategy require organisations to describe their resilience, taking into account different climate scenarios, including the 2° or lower scenario.

Potential benefits due to the implementation of the recommendations are described as follows:

- facilitate access to capital by increasing the confidence of investors and creditors that climate risks are adequately monitored and managed;
- satisfy more effectively the current requests for the communication of material information in financial documents;
- increase awareness and understanding of risks and opportunities related to climate within the company resulting in better risk management and more informed strategic planning; and
- proactively address the growing investor demand for climate information, which could ultimately reduce the number of these requests.

5.6. INNOVATIVE FINANCIAL MECHANISMS: GREEN AND SUSTAINABLE FINANCE

The great attention paid to the importance of ESG factors involves a better evaluation of those companies which:

- are able to mitigate the current climate change;
- pursue decarbonisation strategies;
- adopt a low use of fossil fuels;
- pay attention to renewable energies; and
- are not the cause of high GHG emissions whose effect is to overheat the Earth's surface.

In brief terms, these companies are managed in a responsible manner and pay attention to environmental sustainability without renouncing financial performance.

Sustainability has always been the crucial topic of the European Commission's initiatives, as the EU is committed to promoting development that meets the current needs without compromising the ability of future generations to meet their needs.

European Commission: Projects on Sustainable Finance:

- *Communication from the commission to the European Parliament, the council, the European economic and social committee and the committee of the regions. Next steps for a sustainable European future European action for sustainability*, Strasbourg, 22 November 2016 COM(2016) 739 final, https://ec.europa.eu/europeaid/sites/devco/files/communication-next-steps-sustainable-europe-20161122_en.pdf.

- *Commission staff working document. Key European action supporting the 2030 Agenda and the Sustainable Development Goals*, Strasbourg, 22 November 2016 SWD(2016) 390, https://ec.europa.eu/europeaid/sites/devco/files/swd-key-european-actions-2030-agenda-sdgs-390-20161122_en.pdf

- High-level Expert Group on sustainable finance (HLEG), *Final report 2018 – Financing a sustainable European Economy*, 31 January 2018, https://ec.europa.eu/info/sites/info/files/180131-sustainable-finance-final-report_en.pdf

- *Action Plan – Financing Sustainable Growth*, 8 March 2018, http://eur-lex.europa.eu/legal-content/EN/TXT/PDF/?uri=CELEX:52018DC0097&from=EN

- EC, *High level conference on sustainable finance*, Brussels, 22 March 2018.

EU Commission Regulations 24 May 2018

On 24 May 2018, the European Commission presented a package of three regulations aimed at implementing several key initiatives announced in March 2018 with the *Financing Sustainable Growth* Action Plan to be adopted by the European Parliament and the European Council: (i) the definition of a framework to facilitate sustainable investments, information on sustainable investments and sustainability risks; (ii) the partial modification of directive (EU) 2016/2341 (IORP2); and (iii) the integration of regulation (EU) 2016/1011 with two new low-carbon benchmark parameters and positive carbon impact.

These regulation proposals are part of a wider commission initiative on sustainable development which places ESG issues at the heart of the financial system to support the transformation of the European economy into a greener, more resilient and circular system.

1. *Proposal for a regulation of the European Parliament and of the Council on the establishment of a framework to facilitate sustainable investments.*

   The proposed regulation COM (2018) 353 defines the conditions and framework for gradually creating a unified classification system ('taxonomy') of environmentally sustainable economic activities. The definition of a taxonomy is considered a first and essential step to direct investments in sustainable activities.

   To make investments more sustainable, ESG factors should be considered in the investment decision-making process, integrating, for example, information on GHG emissions, depletion of resources or working conditions. The present proposal and the annexed legislative acts therefore seek to integrate the ESG considerations into the investment and advisory process in a coherent manner across the various sectors.

In particular, this proposal establishes uniform criteria for determining whether an economic activity is environmentally sustainable and establishes a process that promotes the creation of a multi-stakeholder platform to define a homogeneous EU classification system based on a set of criteria specific, in order to determine which economic activities are considered sustainable.

2. *Proposal for a regulation on information on sustainable investments and sustainability risks and amendment of directive (EU) 2016/2341.*

The proposed regulation COM (2018) 354 indicates that to make investments more sustainable, ESG factors should be considered in the investment decision-making process and this proposal and the attached legislative acts aim to integrate the ESG considerations into the investment and consulting process in a coherent way in the various sectors.

This proposed regulation aims to introduce disclosure requirements on how institutional investors and asset managers integrate ESG factors into their risk assessment processes. Requirements for integrating ESG factors into investment decision-making processes, as part of their duties towards investors and beneficiaries, will be further specified through delegated acts.

3. *Proposal for a regulation amending regulation (EU) 2016/1011 on low-carbon benchmark parameters and positive impact on carbon.*

The proposed regulation COM (2018) 355 indicates that ESG factors should be considered in the investment decision-making process to make investments more sustainable, also taking into account GHG emissions for assessment of which different categories of indices have been proposed as low-carbon content.

A growing number of investors pursue investment strategies in low-carbon assets and use specific benchmarks to refer to or measure the performance of their portfolios. While some benchmarks aim to reduce the overall carbon footprint of a standard investment portfolio, others aim to select only components that contribute to achieving the 2 °C target set in the Paris Climate Agreement. Despite the differences in terms of objectives and strategies, all these benchmarks are commonly accepted as a low-carbon benchmark.

Divergent approaches to benchmarking methods lead to fragmentation of the internal market as benchmark users do not have a clear perception that a particular low-carbon index can be considered a benchmark aligned to the 2 °C objective or simply a parameter which aims to reduce the carbon footprint of a standard investment portfolio.

It is therefore necessary to introduce a clear distinction between low-carbon impact benchmarks and those with a positive carbon impact. While the underlying assets referring to a low-carbon benchmark should be selected with the aim of reducing the carbon emissions of the portfolio index relative to the parent index, a positive carbon impact index should include only components whose emissions savings exceed their carbon emissions.

Each company whose assets are selected with underlying assets compared to a positive impact benchmark should save more carbon emissions than it produces and this has a positive impact on the environment. Asset and portfolio managers who claim to pursue an investment strategy compatible with the Paris Climate Agreement should therefore use positive carbon impact benchmarks.

Article 3: Paragraph 1 introduces points 23a and 23b into Regulation (EU) 2016/1011 for the definition of two new categories of benchmark parameters:

(23a) 'low-carbon benchmark' means a benchmark where the underlying assets, for the purposes of point 1(b)(ii) of this paragraph, are selected so that the resulting benchmark portfolio has less carbon emissions when compared to the assets that comprise a standard capital-weighted benchmark and which is constructed in accordance with the standards laid down in the delegated acts referred to in Article 19a and

(23b) 'positive carbon impact benchmark' means a benchmark where the underlying assets, for the purposes of point 1(b)(ii) of this paragraph, are selected on the basis that their carbon emission savings exceed the asset's carbon footprint and which is constructed in accordance with the standards laid down in the delegated acts referred to in Article 19a.

5.7. ESG AND SUSTAINABLE FINANCE

Clear indicators of ESG topics has therefore begun to be requested by institutional investors who increasingly take into account the integration of ESG metrics to jointly evaluate the value creation of companies and sustainability. (London Stock Exchange Group, 2017)

Table 3 presents the 30 indicators of the ESG Guidance and Metrics by the World Federation of Exchanges (WFE), as an example of the type of information available on the level

Table 3. ESG Issues.

Environmental Metrics	Social Metrics	Governance Metrics
- Emissions GHG	- CEO pay ratio	- Board diversity
- Emissions intensity	- Gender pay ratio	- Board Independence
- Energy usage	- Employee turnover	- Incentivised pay
- Energy intensity	- Gender diversity	- Collective bargaining
- Energy mix	- Temporary worker ratio	- Supplier code of conduct
- Water usage	- Non-discrimination	- Ethics and anti-corruption
- Environmental operations	- Injury rate	- Data privacy
- Environmental oversight	- Global health and safety	- Sustainability reporting
- Environmental policy	- Child and forced labor	- Disclosure practices
- Climate risk mitigation	- Human right	- External assurance

Source: WFE. Thirty indicators of ESG metrics processing by the authors.

of ESG risks that these indicators aim to measure, but also about the opportunities, they can derive from sustainability policies in these areas.

As Larry Fink, CEO of BlackRock clarifies: whenever ESG issues are well managed, this is often a signal of operational excellence.

Larry Fink's (2016) Corporate Governance Letter to CEOs states:

> *Generating sustainable returns over time requires a sharper focus not only on governance, but also on environmental and social factors facing companies today. These issues offer both risks and*

> *opportunities, but for too long, companies have not considered them core to their business – even when the world's political leaders are increasingly focused on them, as demonstrated by the Paris Climate Accord. Over the long-term, environmental, social and governance (ESG) issues – ranging from climate change to diversity to board effectiveness – have real and quantifiable financial impacts. (https://www.blackrock.com/corporate/investor-relations/larry-fink-ceo-letter)*

Companies which are responding to the multiple global challenges imposed by sustainability through the search for new solutions to increase financial performance and offer long-term competitive advantages seek to mitigate potential risks through the integration of sustainability in business to protect the value of the brand and guarantee a stable demand for their products over the long term (BlackRock, 2017).

However, those companies that perform poorly on ESG factors are exposed to financial risks: they incur future legal disputes that can affect the company's reputation (NN Investment Partners and European Centre for Corporate Engagements, 'The materiality of ESG factors for equity investment decisions: academic evidence', '… Excluding firms with controversial behavior from the universe has helped improve performance in the research period past years. We find that returns improved further when not only "severe" and "high" controversies but also "significant" controversies are excluded' 2016), bear higher costs for obligations (Karpoff, Lott, & Wehrly, 2005; Orlitzky & Benjamin, 2001) imposed by new regulations, are subject to environmental disasters and risk losing competitive advantage as against more innovative competitors (McWilliams & Siegel, 2001).

Good 'corporate citizenship' often brings with it a lower turnover of employees, excellent brand reputation and customer loyalty; it creates a more committed workforce which in turn improves financial performance.

For the purposes of this work, there are many terms associated with strategies that integrate ESG issues such as – SRI, sustainable and responsible investment, ESG investment, extra financial analysis, corporate social responsibility, ethical finance and non-financial information, which are considered interchangeable.[2] In this section, they have been merged under the heading 'sustainable investment', meaning an investment process that integrates traditional financial analysis with non-financial ESG factors aimed at improving long-term performance and risk management in investment portfolios (PRI, 2015, 2016, 2017, 2020; World Business Council for Sustainable Development, 2017).

Methodologies for Interpreting Sustainable Finance

A number of interesting documents are useful in explaining and discussing sustainable finance, such as:

(1) the one published by CFA Institute (2015, October) entitled *Environmental, Social and Governance Issues in Investing. A Guide for Investment Professionals*;

(2) a study published by Deutsche Asset & Wealth Management (2015, December), *ESG & Corporate Financial Performance: Mapping the Global Landscape*;

(3) a report by EFAMA (2016, September), *Report on Responsible Investment*;

(4) EUROSIF (2016), *European SRI Study 2016*;

(5) Merz Thomas (2016, May), Investment & Pensions Europe, *Sustainable Investing Is Becoming Much More Important*;

(6) Sustainable Stock Exchanges Initiative (2016), *Model Guidance on Reporting ESG Information to Investors*; and

(7) AP7 (2011), *The Performance of Socially Responsible Investment. A Review of Scholarly Studies Published 2008–2010.*

The methodology used to interpret sustainable finance is not unique; there are different types described as follows:

(1) *Exclusionary screening*: In order to limit potential reputational risks, investors may decide to exclude companies and sectors that cause high risks from the investable universe of their portfolio. Screening for exclusion aims to avoid investing in securities of companies or countries and sectors based on traditional moral values (e.g., products or services that include alcohol, tobacco, gambling, nuclear, pornography, animal testing, money laundering, have LGBT – Lesbian, Gay, Bisexual and Transgender – lifestyles), standards and laws (e.g., those relating to human rights and environmental protection). In the process of exclusion based on moral values, attention is focussed on the activity of society and all sectors are excluded; in the process of norm-based screening, the focus is on the behaviour of society in relation to accepted international standards in areas such as human rights and labour standards (ILO, UN Global Compact and OECD guidelines). Where the

consideration of these values is also incorporated in the current legislation, for example, a prohibition on the financing of controversial weapons, exclusive screening can also become a legal obligation.

The exclusion process typically includes an assessment of the level of revenues or other profits that are generated by the security that could be excluded.

(2) *Best in class selection*: The best in class approach involves choosing companies with the best ESG performance compared to the sector peers regardless of the sector they belong to; it can be implemented taking into account the level and changes in ESG performance, that is to say invest more in companies with better levels of ESG performance or 'momentum' compared to industry peers. The best in class methodology is sometimes defined as positive selection or positive alignment, and other approaches that fall into the same category are indicated by the terms best-in-universe and best-effort.

(3) *Engagement and active ownership*: This is a practice of dialogue with companies on ESG issues and the exercise of participation rights aimed at improving the performance of companies. Engagement with a company could be to monitor or influence the results according to ESG practices. Active share ownership initiatives are in sharp contrast to the idea that investors should simply alienate investments in companies with questionable practices. Activism varies in terms of aggressiveness of the approach; some investors may use publicised and contrasting measures, while others may prefer a more discrete approach.

The following are some of the main initiatives indicated by various authors. They are part of an active engagement and shareholder strategy:

- sending a communication to the company indicating the issues on which it is intended to start a comparison;
- request for meetings with company representatives;
- raise a motion or resolution at a general meeting of shareholders;
- vote in the general meetings of shareholders;
- trying to get a seat on the board;
- request to attend an extraordinary/special meeting with shareholders;
- enter a complaint with the regulator/authority; and
- publish a press release on the media.

Engagement and proxy voting practices on sustainability issues are defined as Stewardship. (In 2016, the Italian Association of Asset Management, that is, Assogestioni, has published the Italian Stewardship principles for the exercise of administrative and voting rights in listed companies.) The achievement of the results desired by the active shareholding initiatives takes time and is not exempt from costs; some investors prefer to share their resources and outsource certain activities related to engagement initiatives.

(4) *Thematic investment*: This refers to investments that consider, for example, social, industrial and demographic impacts. A number of investment objectives take account of ESG aspects, including sustainable technologies, real estate, forestry, agriculture, education and health, water and CO_2 emissions.

(5) *Impact investing*: This refers to an investment strategy pursued with the declared intention of generating social and environmental benefits jointly with the achievement of positive financial performance. According to the Global Impact Investing Network, an impact investment practice has four fundamental characteristics:

- the investor's intention to generate a positive social or environmental impact through investment is essential for influencing an impact investing strategy;

- investments with expectations of a return: impact investments are expected to generate a financial return on the invested capital or, at a minimum, a return on capital;

- asset class: impact investments classify financial returns from below the market at risk-adjusted market rates and can be made by considering different asset classes, including, but not limited to, cash, fixed income securities, risk capital and private equity; and

- impact measurement: a hallmark of the impact investment is the investor's commitment to measure and report on social and environmental benefits and the progress of the underlying investments, ensuring transparency and accountability, communicating the investment and the sector.

(6) *ESG integration*: The ESG factor integration strategy at each stage of the investment process refers to a systematic and explicit integration of ESG risks and opportunities into investment analysis. Unlike the best in class method, the integration of ESG factors does not necessarily require benchmarking with peers or an overlap (underestimation) of leaders (laggards).

5.8. MOBILISING GREEN AND SUSTAINABLE FINANCE

How financial centres mobilising capital to deliver key environmental, climate and sustainable development targets?
Milan: Financial Center for Sustainability

Milan is Italy's main financial centre (financial centres are cities with an intense concentration of financial activity involving an interlocking set of financial sectors and transactions); a key challenge is how to improve access to green finance for the country's four million SMEs (G20. Green finance Study Group, 2016). The engagement of Italy's financial sector with environmental drivers of value (such as climate change) has risen in attention, notably through the recent National Dialogue on Sustainable Finance (Forum per la Finanza Sostenibile, ABI, & ANIA, 2016).

On 23 January 2014, the United Nations Environment Programme (UNEP) *Inquiry into the Design of a Sustainable Financial System* was launched in Davos. Its purpose is to explore how the financial system can be aligned with sustainable development; specific focus is on the environmental aspects and the approach is strongly oriented to the analysis of national best practices and to the identification of the most suitable improvement options, taking into consideration the specificities of the individual countries.

Pursuing sustainable development is a strategic challenge for all the financial centres of the world as places where finance supply and demand meet, and where actions to connect finance and sustainability become real thanks to the initiative of the banks, capital markets, insurance companies and institutional investments.

In February 2016, UNEP and the Ministry for the Environment and Protection of the Territory and the Sea started a National Dialogue on Sustainable Finance, also Italy, involving the institutions and the main private subjects of the banking sectors, insurance and asset management (investors), on a broad programme of activities, with the aim of drafting a country report *Financing the Future: Report of Italy's National Dialogue on Sustainable Finance* on the potentials and options of intervention, presented on 6 February 2017 at the Bank of Italy.

Harnessing the financial system will be essential to make a successful transition to a low-carbon, inclusive and sustainable model of development, which regenerates natural capital. Finance lies at the heart of the key international policy achievements reached in 2015: a new set of 17 SDGs and the Paris Agreement on climate change.

The centrepiece of the 2030 Agenda for Sustainable Development are 17 SDGs bringing together an interlocking set of economic, social and environmental objectives with targets through to 2030, matched by 169 targets. For the financial system, the SDGs set out a high-level roadmap for generating 'shared value' – shifting capital away from damaging 'business as usual' trends and towards an end to poverty, increased prosperity with social inclusion and environmental regeneration. Estimates suggest that US$5–7 trillion a year is needed to implement the SDGs globally (UNCTAD, 2014, World Investment Report 2014: Investing in the SDGs).

The Paris Agreement agreed to 'making financial flows consistent with a pathway towards low greenhouse gas emissions and climate-resilient development'.[3] As indicated, The Paris Agreement means aligning capital with the long-term goal of keeping global warming 'well below 2 °C above pre-industrial levels', with the aspiration to 'limit temperature increase to 1.5 °C'. It also gave a higher profile to financing adaptation to growing climate shocks.

In June 2017, the UN Environment organisation within the G7 environment programme prepared a *Financial Centers for Sustainability* report to explore how the financial centres are able to contribute to the implementation of the 17 SDGs indicated in the UN Agenda 2030 and the 2015 Paris climate agreements (COP21).

To this end, an international network of Sustainability Financial Centres was created to share experiences and undertake actions on common priorities. On 28 September 2017, the first global meeting promoted by the Casablanca Finance City Authority was hosted in Casablanca, Morocco. UN Environment, in collaboration with the Italian Ministry of the Environment and Morocco's Presidency of the Climate Conference held in 2016 (COP22), attended this meeting.

A total of 14 financial centres (Astana, Casablanca, Dublin, Hong Kong, London, Luxemburg, Milan, Paris, Qatar, Shanghai, Stockholm, Tokyo, Toronto and Zurich) joined the conference in Casablanca. Sara Lovisolo of Borsa Italiana co-chaired the Italian Sustainable Finance Initiative Working Group.

During the conference, a joint declaration was signed entitled the *Casablanca statement on financial centers for sustainability* whose main objectives are indicated as follows:

- promote strategic actions in the respective financial centres that have as their object the green and sustainable finance;

- share knowledge to broaden human skills and measure the contribution of the financial centre to climate action and sustainable development;

- work together to expand 'green assets and products';

- work with political leaders of cities, regional, national and international levels to achieve positive conditions for green and sustainable finance; and

- promote an international network of Sustainability Financial Centers and help them for all operational aspects.

NOTES

1. For example, some recommendations by Global Task Force members in 2016–2019 are as follows: C40 Call for Action on Municipal Infrastructure and Finance (October 2016), Bonn-Fiji Commitment of Local and Regional Leaders to Deliver the Paris Agreement at All Levels (November 2017), GCAS Call to Action (September 2018), IPCC SR 1.5 (October 2018), Summary for Urban Policymakers: What does IPCC SR 1.5 Mean for Cities? (December 2018), Cities and Regions Talanoa Dialogues: Leveraging Subnational Action to Raise Ambition (December 2018) and From Talanoa Dialogue to NDCs: Shifting Climate Ambition through Multilevel Action (April 2019).

2. For more details Morgan Stanley Institute for Sustainable Investing, 2015a, 2015b.

3. UNFCCC (2015). The Paris Agreement, Article 2c: http://unfccc.int/files/home/application/pdf/paris_agreement.pdf.

6

MONITORING, EVALUATION AND REPORTING ON SDG 13 IMPLEMENTATION

6.1. NEW TRENDS IN CORPORATE REPORTING

Business leaders should identify essential resources and related threats which have impact on running their businesses, while looking at and accounting for the environmental and social impacts of corporate activities. These practices are increasingly considered as strategic tools that allow CEOs and a company's management to learn about risks related to the external and internal environment, and to design tailor-made strategies for adapting to those potential threats and turn them into opportunities (Accenture Sustainability Services, 2013). Furthermore, stakeholders are increasingly focussing on companies' behaviour and features, such as reputation, commitment to social issues (i.e., gender diversity inclusion, quality of employment and so on), transparency of business investments and operations. Investors increasingly seek projects and products which observe ESG standards . Hence, greater attention has been given to private sector behaviour, as well as how well business strategies embed sustainability issues in their corporate business, or how they communicate and which information they make available to stakeholders. This is an important challenge for the image of the private sector. There is a widespread recognition of the importance of business in contributing to labour market dynamics, in improving and developing technologies and in rising national income by paying taxes, for example. It is time to look at the nature and the purpose of business from a long-term perspective, so that business activities can contribute to more sustainable and inclusive growth.

Therefore, the government's role will be essential in creating the right framework, within which companies are supported in integrating concepts of sustainability into their business and reporting cycle. Disclosing information about a company's social responsibility is not yet a mandatory practice in several countries. However, stakeholders are beginning to require businesses to go beyond their economic and legal responsibilities and contribute to the improvement of all stakeholders.

6.2. FROM VOLUNTARY TO COMPULSORY DISCLOSURES

Recent developments in the field of corporate reporting at an international level are making significant changes to a company's information systems, in particular as regards the preparation of the annual report and other related reports. The strong impetus towards a real increase in the production of non-financial information has led to the creation of a new scenario in corporate reporting, whereby recently it has been noted that the data on the environment, health and safety of workers and corporate governance are increasingly disclosed in annual reports, in integrated reports or on dedicated web pages, rather than being included exclusively in stand-alone reports.[1]

The contents have also changed, moving from a perspective mainly focussed on environmental aspects (emissions and discharge, waste management) to an approach more specifically focussed on social issues such as health and safety at work (injuries and fatalities) or issues that reflect the expectations of a broad group of stakeholders such as a sustainable supply chain, gender diversity, ethics and respect for human rights. Considering the current trend in favour of a progressive improvement of information focussed on sustainability and on climate risks, between 2016 and 2017 many countries, including Austria, Belgium, Canada, Denmark, Czech Republic, Finland, France, Germany, Greece, Hungary, India, Italy, the Netherlands, Panama, Romania, the UK and the USA, have introduced regulations that require ESG information supported by specific guidelines. Recently an evolution is underway from a voluntary disclosure of non-financial information to a communication that is becoming compulsory and regulated by law. Some recent initiatives are described as follows:

- The Council Directive (EU) 2016/234 of the European Parliament (IORP2)1 requires European pension funds to publish a report which explains, among the principles of investment policy, which ESG factors are integrated into finance investment decisions and the risk control and management system related to climate change.

  An even more ambitious initiative in terms of environmental reporting has been French legislation of Article 173-VI of the *LOI n° 2015-992 du 17 août 2015 relative à la transition énergétique pour la croissance verte*. Pension funds, insurance companies and asset owners are required to disclose in the annual report information useful for understanding how environmental considerations are taken into account in investment decisions. They must explain the resources used to contribute to the energy and ecological transition and describe the climate risk exposure, including the measurement of GHG emissions associated with their investment portfolio (carbon footprint, carbon intensity and carbon efficiency). Moreover, they have to indicate the contributions given to the international objective of limiting climate heating (http://www.eurosif.org/wp-content/uploads/2014/07/coporate-pensiosn-funds.pdf).

- Legislative Decree. No. 254/2016 published in Italy in the Official Gazette no. 7 of 10 January 2017 required large companies to prepare a non-financial statement in addition to the annual financial statement for the fiscal year 2017. The companies are characterised by an average number of more than 500 employees and exceeding either a total balance sheet of EUR 20 million or a net turnover of EUR 40 million. This non-financial statement must include, at a minimum, information on the use of energy resources, distinguishing between those produced from renewable and non-renewable sources, the use of water resources, GHG emissions and pollutant emissions in the atmosphere, social and employee-related aspects and human rights, anti-corruption and bribery matters (https://www.gazzettaufficiale.it/eli/id/2017/01/10/17G00002/sg).

With respect to sustainability, therefore, in the last decade there has been an increase of indicators that can be judged to be 'disproportionate' compared to the effective needs of company's business activities. The fundamental problem therefore lies in the adequate choice of reporting standards and the related ad hoc guidelines through which companies communicate their choices and policies regarding the implementation of the SDGs.

Particularly with reference to Goal 13 'Climate action', it is interesting to investigate the indicators related to climate change disclosure that are stimulating extensive academic and professional literature with in-depth analysis in terms of review papers and empirical studies (Gusmao Caiado et al., 2018; Leal Filho et al., 2017).

6.3. SDGs AND GLOBAL REPORTING INITIATIVE

In the international context, the Global Reporting Initiative (GRI) plays a crucial role in the scenario of sustainability reporting; particularly on SDGs reporting, it has provided specific documents in order to facilitate the alignment of communication of economic, social and environmental aspects (triple bottom line). The GRI, in coordination with two other important organisations, the United Nations Global Compact (UNGC)[2] and the World Business Council for Sustainable Development (WBCSD),[3] has prepared a document (SDG Compass Guide[4]) to explain some operational aspects in particular of their reporting disclosure on *How the SDGs affect your business – offering you the tools and knowledge to put sustainability at the heart of your strategy*.

Despite this, an effective alignment, in particular with the GRI G4/GRI standards and the UNGC principles, and a detailed indication on how such information should be communicated and represented in corporate reporting, are still needed.[5]

To address this need, the SGDs Compass Guide describes five steps that companies should follow in order to effectively communicate the SDGs (http://sdgcompass.org/):

(1) The first step concerns the understanding of goals, 'understanding the SDGs'.

(2) The second step 'defining priorities' involves the definition of priorities or careful consideration by companies of current or potential impacts, both in positive and negative terms, that the SDGs can have on business opportunities and in general on the value chain.

(3) The third step 'setting goal' consists in explaining the objectives as critical for the company's success that must be appropriately 'aligned' with the SDGs so that the company leadership is able to declare their commitment to sustainable growth.

(4) The fourth term 'integrating' points out the need to integrate the SDGs in the core business and in corporate governance in order to root the goals within the company's functions. It may be useful to establish partnerships with other companies, that is, companies in the same sector, or with governments or civil society organisations.

(5) The last and most important step, that is, 'reporting and communicating', concerns reporting and communication of the SDGs. The achievement of goals allows companies to communicate the priorities in terms of performance related to sustainable development, and this issue has to be harmonised within corporate reporting by adequate indicators. The continuous communication of progress on achieving the goals must also be in line with stakeholders' expectations.

In general terms, an 'effective communication' of the SDGs can be implemented in different ways: for example, companies can choose whether to produce an ad hoc report or a 'stand-alone' report for the SDGs or use a document where SDGs can be emphasised graphically by inserting icons. Alternatively, SDGs can be added to a table of contents, such as the GRI Report Index, where a column can be added to highlight the correlation between the main GRI indicators and the SDGs. With reference to each goal, the guide also offers indications regarding the role played by business activities, the 'key business themes' that are specifically affected by the goal, provides a detailed list of examples of 'key business actions and solutions', examples of 'key business indicators' and 'key business tools' and finally the details about the SDGs' targets.

With reference to SDG 13, the most important indicators (see Table 4) have been identified in four GRI G4 indexes on gas emissions (G4 EN 15, G4 EN 16, G4 EN 17 and G4 EN 18). There is also an additional indicator suggested by Global Compact about involvement in promoting resilient practices or improving procedures in the value chain to respect climate change.

Moreover, carbon emissions reveal an important contribution in the business context, particularly in some economic sectors, such as industry, transportation and electricity. Several studies analysed carbon emissions and related carbon accounting in order to support companies in recording their GHG emissions and monitoring their GHG emissions and plan mitigation actions. For example, some scholars identified a guide for carrying out a GHG emissions inventory which enables them to use their activity data to produce the organisation's total emissions. The total emissions will be broken down based on indicator, source and scope. This would enable industries to identify the highest contributor of emissions either by indicator or source, and to identify which scope their emissions fall under. The carbon emissions profile and carbon emissions index can be used to identify and analyse the organisation's emissions performance (Jusoh & Hashim, 2018).

Table 4. Business Theme GHG Emission.

Type of Indicator	Source of the Indicator	Indicator Description	Indicator ID
General	Carbon Disclosure Project (CDP's) 2015 Climate Change Information Request	How do your gross global emissions (Scopes 1 and 2 combined) for the reporting year compared to the previous year?	CC12.1
General	CDP's 2015 Climate Change Information Request	Please describe your gross global combined Scope 1 and 2 emissions for the reporting year in metric tonnes CO_2e per unit currency total revenue	CC12.2
General	CDP's 2015 Climate Change Information Request	Please describe your gross global combined Scope 1 and 2 emissions for the reporting year in metric tonnes CO_2e per full time equivalent employee	CC12.3
General	CDP's 2015 Climate Change Information Request	Please account for your organization's Scope 3 emissions, disclosing and explaining any exclusions	CC14.1
General	CDP's 2015 Climate Change Information Request	Are you able to compare your Scope 3 emissions for the reporting year with those for the previous year for any sources?	CC14.3
General	CDP's 2015 Climate Change Information Request	Does the use of your goods and/or services directly enable GHG emissions to be avoided by a third party?	CC3.2

Table 4. (Continued)

Type of Indicator	Source of the Indicator	Indicator Description	Indicator ID
General	CDP's 2015 Climate Change Information Request	Please identify the total number of projects at each stage of development, and for those in the implementation stages, the estimated CO_2e savings	CC3.3a
General	CDP's 2015 Climate Change Information Request	Please provide your gross global Scope 1 emissions figures in metric tonnes CO_2e	CC8.2
General	CDP's 2015 Climate Change Information Request	Please provide your gross global Scope 2 emissions figures in metric tonnes CO_2e	CC8.3
Sector-specific	GRI G4 Construction and Real Estate Sector Disclosures	Greenhouse gas emissions intensity from buildings	CRE3
Sector-specific	GRI G4 Construction and Real Estate Sector Disclosures	Greenhouse gas emissions intensity from new construction and redevelopment activity	CRE4
Sector-specific	GRI G4 Electric Utilities Sector Disclosures	Allocation of CO_2e emissions allowances or equivalent, broken down by carbon trading framework	EU5
General	GRI G4 Sustainability Reporting Guidelines	Direct GHG emissions (Scope 1)	G4-EN15
General	GRI G4 Sustainability Reporting Guidelines	Energy indirect GHG emissions (Scope 2)	G4-EN16
General	GRI G4 Sustainability Reporting Guidelines	Other indirect GHG emissions (Scope 3)	G4-EN17
General	GRI G4 Sustainability Reporting Guidelines	GHG emissions intensity	G4-EN18

Table 4. (Continued)

Type of Indicator	Source of the Indicator	Indicator Description	Indicator ID
General	GRI G4 Sustainability Reporting Guidelines	Reduction of GHG emissions	G4-EN19
General	GRI G4 Sustainability Reporting Guidelines	Extent of impact mitigation of environment impacts of product and services	G4-EN27
General	GRI G4 Sustainability Reporting Guidelines	Significant environmental impacts of transporting products and other goods and materials for the organization's operations, and transporting members of the workforce	G4-EN30

Source: SDG Compass Guide – http://sdgcompass.org/business-indicators/?filter_sdg_goal=13.

Some operational tools that can be used to achieve the various targets such as the Green House Protocol (http://www.ghgprotocol.org/), the Impact Reporting and Investment Standards (https://iris.thegiin.org/) and the Building a Resilient in Power Sector (http://bcsdh.hu/wp-content/uploads/2014/04/WBCSD_Building-a-Resilient-Power-Sector-Interactive.pdf) are mentioned by GRI.

6.4. SDGs AND INTEGRATED REPORTING

The SDGs and Agenda 2030 represent global multi-stakeholder responses to the challenge of sustainable development. Given the interdependency of the SDGs and the involvement in trade-offs, it is possible to identify an alignment with the business's outcomes and multiple capitals framework

(ICAS & IIRC, 2017). The report, published in 2017 on SDGs and integrated reporting (IR), identified five steps for contributing to the SDGs through IR (Adams, 2017, p. 7). This five-step process includes:

(1) Understand sustainable development issues relevant to the organisation's external environment.

(2) Identify material sustainable development issues that influence value creation.

(3) Develop integrated thinking connectivity and governance.

(4) Prepare the integrated report.

This document does not develop indicators or prioritised disclosures to report contributions to the SDGs. Rather it discusses the concepts, guiding principles and content elements of the IR framework and its notion of integrated thinking to help organisations respond to the SDGs. One of the most interesting parts is focussed on the alignment of SDGs and the IR value creation process set out in the IR framework

In the figure disclosed in the IIRC document (Adams, 2017) related to the representation of IR business model, it is possible to see the link of SDG 13 with all six capitals categorised by the IIRC framework. Climate change represents an important global trend that can be analysed in connection with the transformation of the six capitals in relation to one or more SDGs. For example, increased reliance on renewable energy sources and improving diversity in the work force enhance natural and human capital and may contribute to the achievement of SDGs 5, 7, 10 and 13. In summary, financial capital can be related to 14 of the goals, manufactured capital to 10 of the goals, intellectual capital to 9 of the goals, human capital to 12 of the goals, social and relationships to all 17 of the goals and natural capital to 9 of the goals (Adams, 2017, p. 22).

NOTES

1. Corporateregister.com, 2013.
2. https://www.unglobalcompact.org/
3. https://www.wbcsd.org/
4. https://sdgcompass.org/
5. See also Sustainability Accounting Standars Board, 2016.

7

CONCLUSIONS

This work has analysed some important factors about the SDGs, particularly focussing on SDG 13 and climate change. The SDGs represent an ambitious and transformative agenda that involves many actors, at a global and national level, in the public and private sectors, with a special role played by sustainable finance. In several documents, SDG 13 seems to be one of the most investigated goals, being, as it is, concerned with a crucial global trend, that is, climate change.

Countries, governments, international associations, public and private sectors and businesses can contribute to the achievement of these ambitious goals. However, the potential of the SGDs to guide governments, communities and organisations towards a shared and lasting process of sustainable development reveals a significant role (Hajer et al., 2015). Among numerous actors, businesses play a vital role in achieving the SDGs that, at the same time, can serve as a useful tool for companies to measure their performance.

Taking the same view, academic literature confirms this aspect since interdisciplinary SDG studies include the fields of business management (Annan-Diab & Molinari, 2017;

Schaltegger et al., 2017; Storey et al., 2017) and accounting/accountability (Bebbington et al., 2017; Bebbington & Unerman, 2018). In addition, some studies highlight the importance of a number of external factors in the process of SDG implementation, such as country-level factors, institutions and internal factors such as structural and organisational characteristics (company size, endowment of resources, level of skills and intangibles, level of commitment and sustainability performance and presence of external assurance), which together with individual motivations (personal variables) influence the companies' choice to report on SDGs.

Other interesting analyses are being carried out to identify the most treated themes/objectives (Salvia et al., 2019) in the approach of a sample of experts (266) belonging to different countries (Table 2: Number and percentage of experts according to their position, p. 845.), highlighting the correlation between specific problems afflicting the individual areas of the planet and the interest/priorities of orientation (in Africa, e.g., the themes of poverty, access to water and sanitary resources prevail; in Europe education, industry, infrastructural innovation are priorities). Alongside 'regional' differences, some objectives, such as climate change, have global reach and stimulate the attention of the scientific community from an interdisciplinary perspective. The relationship between local problems and global challenges constitutes a research front whose implications are important for the implementation on a regional basis of the SDGs, as well as deepening comparative analysis into problems that limit the spread of some goals and the resources available in different countries. For example, in Table 5 it is possible to note that SDG 13 has been widely investigated in different studies.

Table 5. SDGs Most and Least-Researched Globally.

SDGs most researched	4, 6, 11, 12, 13 and 15
SDGs least researched	1, 2, 3, 5, 7, 8, 9, 10, 14, 16 and 17

Source: Salvia et al. (2019, Table 3, p. 846).

Finally, further studies and assessments are needed to develop both global and country-specific programmes, strategies, investment plans and policies to support all types of organisations in contributing to the successful implementation of the SDGs.

APPENDIX

Table 6. Overview of Definitions.

SDG 13 – Climate Action
Take Urgent Action to Combat Climate Change and Its Impacts

Goal Concepts	Conceptual Proposed Definitions According to Authors	References
Climate change	Climate change is a global change and also a global priority and has been recognised as both one of the biggest threats and biggest opportunities for global health in the twenty-first century (Verner et al., 2016)	Leal Filho et al. (2017)
Climate change vulnerability	Understanding the determinants of adaptive capacity has advanced and confirms the conclusion that developing countries, particularly the least-developed countries, have lesser capacity to adapt than developed countries. This condition contributes to relatively high vulnerability to the damaging effects of climate change	IMPACTS; ADAPTATION; VULNERABILITY, CLIMATE CHANGE 2001:2001
Climate change impacts	What is clear is that climate change impacts are cross-sectoral and multidimensional, and therefore require cross-sectoral mitigation and adaptation approaches.	

Table 6. (Continued)

Targets and Concepts	Conceptual Proposed Definitions According to Authors	References
13.1 – Strengthen resilience and adaptive capacity to climate-related hazards and natural disasters in all countries		
Resilience	Is the ability of a system to absorb external stresses	Haimes (2009)
	Is the capability to create foresight, to recognise, to anticipate and to defend against the changing shape of risk before adverse consequences occur	Haimes (2009)
	Refers to the inherent ability and adaptative responses of a system that enable them to avoid potential losses	Haimes (2009)
	Is the result of a system (1) preventing adverse consequences, (2) minimising consequences and (3) recovering quickly from adverse consequences	Haimes (2009)
	Positive adaptation in the face of adversity	Schoon and Bynner (2003)
Natural hazards	Natural hazards result in significant loss of life and serious economic, environmental and social impacts that greatly retard the development process. Drought differs from other natural hazards (e.g., floods, tropical cyclones and earthquakes) in several ways	Wilhite (2000, chapter 1)

Natural disasters	In simple terms, a natural disaster is a natural event with catastrophic consequences for living things in the vicinity. However, different definitions of natural disasters are often used and some of them are based primarily on loss of life	Sivakumar (2005)
13.2 – Integrate climate change measures into national policies, strategies and planning		
National policies, strategies and planning on climate change mitigation	The impacts of climate change and climate change mitigation policies are socially differentiated, and are therefore matters of local and international distributional equity and justice (Adger, 2001, p. 929; O'Brien et al., 2004; Paavola & Adger, 2006). Some argue that inequality leads to greater environmental degradation and that a more equitable distribution of power and resources would result in improved environmental quality (Agyeman, Bullard, & Evan, 2002; Boyce, Klemer, Templet, & Willis, 1999; Solow, 1991, Stymne & Jackson, 2000)	Jabareen (2013)
13.3 – Improve education, awareness raising and human and institutional capacity on climate change mitigation, adaptation, impact reduction and early warning		
Education on climate change	There is a clear educational agenda in climate change adaptation and mitigation strategies, which requires learning new knowledge and skills and changing behaviours in order to reduce vulnerabilities; manage the risks of climate change; change consumption and production patterns; and build adaptive capacity and resilient societies	Conventions (2012)

Table 6. (Continued)

Targets and Concepts	Conceptual Proposed Definitions According to Authors	References
Climate change adaptation and mitigation	Adaptation responds directly to the impact of increased concentrations of GHGs in both precautionary and reactive ways, rather than through the preventative approach of limiting the source of the gases (this is known as 'mitigation'). This avoids the enormous political obstacles facing initiatives to curtail the burning of fossil fuels by factories, transport and other sectors. Adaptation to climate change is considered especially relevant for developing countries, where societies are already struggling to meet the challenges posed by existing climate variability	Schipper (2016)
Climate change impact reduction	DRR as defined by UNISDR (2004) is 'the systematic development and application of policies, strategies and practices to minimise vulnerabilities, hazards and the unfolding of disaster impacts throughout a society, in the broad context of sustainable development' (p. 3). DRR CCA strategies aim to reduce vulnerability to expected impacts of climate change. However, the concept of CCA is very broad (McGray et al., 2007). CCA strategies exist across local and global scales, from community level responses through to local, national and international government interventions (McGray et al., 2007; UNFCCC, 2006). At the community level, strategies include improvements to agricultural systems such as crop diversification or the introduction of hazard resistant crop varieties; risk assessments and associated plans; the protection of natural resources; early warning systems; education and awareness measures and protection of water resources (UNFCCC, 2006). At the national level for least-developed countries, some countries have developed National Adaptation Programs of Action	Mercer (2010)

Climate change early warning	Early warning can take several forms, ranging from the knowledge that an event could occur, through qualitative assessment that it is becoming more likely, to a forecast of its timing. Recently, there has been growing interest in generic early warning signals for critical transitions in complex systems, especially slowing down as a bifurcation is approached. Furthermore, slowing down has been found in climate-model output and palaeoclimate data approaching abrupt transitions	Lenton (2011)
	13.A – Implement the commitment undertaken by developed-country parties to the United Nations Framework Convention on Climate Change to a goal of mobilising jointly $100 billion annually by 2020 from all sources to address the needs of developing countries in the context of meaningful mitigation actions and transparency on implementation and fully operationalise the Green Climate Fund through its capitalisation as soon as possible	
United Nations Framework Convention on Climate Change	The United Nations Framework Convention on Climate Change (the Convention), adopted on 9 May 1992, was negotiated in response to the growing scientific evidence of the dangers posed by increased concentrations of GHGs (principally carbon dioxide, methane, CFC's and nitrous oxides) in the atmosphere. These could lead to historically unprecedented rates of increase of world average temperatures with consequential adverse effects likely to include sea level rise, increased hurricane and cyclone activity, drought and desertification and coral bleaching	Sands (1992)

Table 6. (Continued)

Targets and Concepts	Conceptual Proposed Definitions According to Authors	References
Developing countries	The United Nations acknowledges that it has 'no established convention for the designation of "developed" and "developing" countries or areas'. According to its so-called M49 standards, published in 1999.	'Composition of macro geographical (continental) region'. United Nations. Archived from the original on 6 March 2010.
	The designations 'developed' and 'developing' are intended for statistical convenience and do not necessarily express a judgement about the stage reached by a particular country or area in the development process.	'Millennium development indicators: World and regional groupings'. UN Statistics Division (2003). Archived from the original on 10 February 2005. Retrieved 13 May 2017.
	The United Nations implies that developing countries are those not on a tightly defined list of developed countries.	'Standard Country and Area Codes Classifications (M49): Developed Regions'. UN Statistics Division. Archived from the original on 11 July 2017. Retrieved 13 May2017.
	There is no established convention for the designation of 'developed' and 'developing' countries or areas in the UN system. In common practice, Japan in Asia, Canada and the United States in northern America, Australia and New Zealand in Oceania, and Europe are considered 'developed' regions or areas. In international trade statistics, the Southern African Customs Union is also treated as a developed region and Israel as a developed country; countries emerging from the former Yugoslavia are treated as developing countries; and countries of eastern Europe and of the Commonwealth of Independent States [the former Soviet Union] in Europe are not included under either developed or developing regions	'United Nations Statistics Division – Standard Country and Area Codes Classifications (M49)': Unstats.un.org. Retrieved 15 January 2014

GCF	The GCF is the newest actor in the multilateral climate finance architecture and became fully operational in 2015, approving US$ 168 million for its first eight projects just weeks before COP21. The GCF is an operating entity of the financial mechanism of the UNFCCC	Schalatek et al. (2015)

13.B – Promote mechanisms for raising capacity for effective climate change-related planning and management in least-developed countries and small island developing states, including focussing on women, youth and local and marginalised communities

SIDS	SIDS are especially vulnerable to climate change and disasters (see Kelman & West, 2009 for a critical review of SIDS and climate change literature). This is due to their shared challenges in sustainable development, geographical location and their subsequent propensity to vulnerabilities and environmental hazards (Lewis, 1999, 2009; McGillivray et al., 2008; Pelling & Uitto, 2001). As a result of these similarities, SIDS have also formed a relatively cohesive group addressing environmental issues including climate change, placing SIDS on the agenda of international policy negotiations (UNFCCC, 2005; CICERO and UNEP/GRID-Arendal, 2008). Anticipated climate change impacts including sea-level rise, increased temperatures, decreased water supplies, increased endemic diseases and deterioration in coastal conditions threaten island populations (Lewis, 1999; O'Shea, 2003; UNFCCC, 2005). Such impacts will undoubtedly affect livelihoods through an increase in, and exacerbation of, hydro-meteorological hazards and changes in seasonal weather patterns	Mercer (2010)

Source: Authors' elaboration.

REFERENCES

Accenture Sustainability Services. (2013). The UN Global Compact – Accenture CEO study on sustainability. Retrieved from https://www.unglobalcompact.org/docs/news_events/8.1/UNGC_Accenture_CEO_Study_2013.pdf

Adams, C. A. (2017).The Sustainable Development Goals, integrated thinking and the integrated report. Retrieved from www.integrated.org

Adger, W. N. (2001). Scales of Governance and Environmental Justice for Adaptation and Mitigation of Climate Change. *Journal of International Development*, *13*(7), 921–931.

Adger, W. N., Benjaminsen, T. A., Brown, K., Svarstad, H. (2001). Advancing a Political Ecology of Global Environmental Discourses. *Development and Change*, *32*(4), 681–715.

Agyeman, J., Bullard, R. D., & Evans, B. (2002). Exploring the Nexus: Bringing Together Sustainability, Environmental Justice and Equity. *Space and Polity*, *6*(1), 77–90.

Anderson A. (2012). Climate Change Education for Mitigation and Adaptation. *Journal of Education for Sustainable Development*, *6*(2), 191–206.

Annan-Diab, F., & Molinari, C. (2017). Interdisciplinarity: practical approach to advancing education for sustainability and for the sustainable developmentgoals. *International Journal of Management Education*, *15*(2), 73–83.

AP7 Sjunde AP-fonden. (2011). The performance of socially responsible investment. *A review of scholarly studies published 2008–2010*. Retrieved from https://www.ap7.se/app/uploads/2017/06/AP7_The-Performance-of-socially-responsible-investment.pdf

Bebbington, J., Russell, S., & Thomson, I. (2017). Accounting and sustainable development: Reflections and propositions. *Critical Perspectives on Accounting, 48,*21–34.

Bebbington, J., & Unerman, J. (2018). Achieving the United Nations Sustainable Development Goals: An enabling role for accounting research. *Accounting, Auditing & Accountability Journal, 31*(1), 2–24.

Bertelsmann Stiftung and Sustainable Development Solutions Network (2018). SDG Index and Dashboards Report, 2018. Global responsibilities. Implementing the goals. Retrieved from http://www.sdgindex.org/assets/files/2018/01%20SDGS%20GLOBAL%20EDITION%20WEB%20V9%20180718.pdf

BlackRock. (2017, March). How BlackRock Investment Stewardship engages on climate risk. Retrieved from https://www.blackrock.com/corporate/en-ae/literature/market-commentary/how-blackrock-investment-stewardship-engages-on-climate-risk-march2017.pdf

Boyce J. K., Klemer, A.R., Templet, P. H., Willis, C. E. (1999). *Ecological Economics, 29*(1), 127–140.

Campbell, B. M., Hansen, J., Rioux, J., Stirling, C. M., Twomlow, S., & Wollenberg, E. L. (2018). Urgent action to combat climate change and its impacts (SDG 13): Transforming agriculture and food systems. *Current Opinion in Environmental Sustainability, 34,* 13–20.

Carbon Tracker Initiative. (2011). Unburnable carbon – Are the world's financial markets carrying a carbon bubble.

Retrieved from http://www.carbontracker.org/wp-content/uploads/2014/09/Unburnable-Carbon-Full-rev2-1.pdf

Carbon Tracker Initiative. (2013). Unburnable carbon Waste capital and stranded assets. Retrieved from http://www.carbontracker.org/wp-content/uploads/2014/09/Unburnable-Carbon-2-Web-Version.pdf

Carbon Tracker Initiative/Climate Disclosure Standards Board. (2016). Considerations for reporting and disclosure in a carbon-constrained world. Retrieved from http://www.carbontracker.org/wp-content/uploads/2016/01/Climate-Risk-Disclosure_CTI-CDSB_v2_single-pages_WEB.pdf

Carbon Trust (2008). Climate change – A business revolution? How tackling climate change could create or destroy company value". Co-analysis with Mc Kinsey & Co. Retrieved from www.carbontrust.com/media/84956/ctc740-climate-change-a-business-revolution.pdf

CFA Institute. (2015, October). Environmental, social and governance issues in investing. A guide for investment professionals. Retrieved from https://www.cfainstitute.org/learning/products/publications/ccb/Pages/ccb.v2015.n11.1.aspx

Church, J. A., & White, N. J. (2006). A 20th century acceleration in global sea level rise. *Geophysical Research Letters*, *33*, L01602, doi: 10.1029/2005GL024826

CICERO and UNEP/GRID-Arendal. (2008). Many Strong Voices: Ouline for an Assessment project design. CICERO Report 2008: 05 Oslo, CICERO. Center for International Climate and Environmental Research, Oslo.

Climate Disclosure Standard Boards (CDBS). (2015). CDSB Framework for reporting environmental information & natural capital. Retrieved from http://www.cdsb.net/sites/

cdsbnet/files/cdsb_framework_for_reporting_environmental_information_natural_capital.pdf

Conventions, T. R. (2012). Climate change education for mitigation and adaptation. UNESCO Special Section on the ESD Response to the Three Rio Conventions. Journal of Education for Sustainable Development, 6(2), 191–206.

Cotter J., & Naja, M. M. (2011). Institutional investor influence on global climate change disclosure practices. *Australian Journal of Management*, 37(2), 169–187.

Cremers, M. (2016). Corporate social responsibility in light of Laudato Si. *The Journal of Corporate Citizenship*, 64, 62–78

DB Climate Change Advisors. (2012, June). Sustainable investing: Establishing long-term value and performance. Retrieved from https://www.dbadvisors.com/content/_media/Sustainable_Investing_2012.pdf

Derksen, C., & Brown, R. (2012). Spring snow cover extent reductions in the 2008-2012 period exceeding climate model projections. *Geophysical Research Letters*, 39, L19504.

Deutsche Asset & Wealth Management. (2015). ESG & corporate financial performance: Mapping the gobal landscape. Retrieved from https://institutional.dws.com/content/_media/K15090_Academic_Insights_UK_EMEA_RZ_Online_151201_Final_(2).pdf

European Fund and Asset Management Association (EFAMA). (2016, September). Report on responsible investment. Retrieved from https://www.efama.org/Publications/Public/Responsible_Investment/140228_Responsible_Investment_Report_online.pdf

EUROSIF. (2016). European SRI study 2016. Retrieved from http://www.eurosif.org/wp-content/uploads/2016/11/SRI-study-2016-HR.pdf

Fink L. (2016). *Larry Fink's 2016 Corporate Governance Letter to CEOs*. Black Rock, Retrieved from https://www.blackrock.com/corporate/en-gb/literature/press-release/2016-larry-fink-ceo-letter.pdf

Forum per la Finanza Sostenibile, ABI, & ANIA. [Forum for Sustainable Finance, ABI & ANIA] (2016, October). Finanza sostenibile e cambiamento climatico. [Sustainable finance and climate change]. Retrieved from https://finanzasostenibile.it/wp-content/uploads/2016/09/Clima-web.pdf

Friede, G., Busch, T., & Bassen, A. (2015). ESG and financial performance aggregated evidence from more than 2000 empirical studies. *Journal of Sustainable Finance & Investment*, 5(4), 210–233.

FSB. (2015a). Proposal for a disclosure task force on climate-related risks. Retrieved from http://www.fsb.org/wp-content/uploads/Disclosure-task-force-on-climate-related-risks.pdf

FSB. (2015b, December 4). FSB to establish Task Force on Climate-related Financial Disclosures. Retrieved from http://www.fsb.org/2015/12/fsb-to-establish-task-force-on-climate-related-financial-disclosures/

G20, Green Finance Study Group. (2016). G20 green finance synthesis report. Retrieved from http://unepinquiry.org/wpcontent/uploads/2016/09/Synthesis_Report_Full_EN.pdf

Gasperini, A. (2013). L'analisi ESG ed i punti di contatto con l'analisi fondamentale (The ESG analysis and common topics with the fundamental analysis). In Dal Maso, D. & Fiorentini, G. (Eds.), *Creare Valore a Lungo Termine – conoscere, promuovere, gestire l'investimento sostenibile e responsabile* [Creation of value in the long term - to know, promote and manage sustainable and responsible investment] (pp. 152–161). Milano: Egea.

Gasperini, A., & Doni, F. (2016). Sustainable Development Goals 13: Quali conseguenze per le aziende? [Sustainable Development Goals: What are the consequences for companies?]. *Amministrazione e Finanza. 31*(12), 10–18

Gasperini, A., & Doni, F. (2017a). *Disclosure of climate risks and ESG information*. AIAF White Paper No. 173. White Paper published by AIAF Association of Italian Financial Analysts, Milan, Italy.

Gasperini, A., & Doni, F. (2017b). Reporting of climate risks and ESG information An investigation on ETF fossil free and low carbon investment. In *18th European roundtable on sustainable consumption and production conference, ERSCP2017*, Skiathos, Greece, 1–5 October.

Gasperini, A., & Doni, F. (2017c). *Bilancio Integrato e di Sostenibilità: la comunicazione dei contributi agli SDGs* [*Integrated reporting and sustainability: Disclosure of contribution to SDGs*). AIAF White Paper No. 173. White Paper published by AIAF Association of Italian Financial Analysts, Milan, Italy.

Gasperini, A., & Doni, F. (2017d). Climate risks disclosure and socially responsible investment strategies. Conference paper presented at the International Conference at the University of Milano-Bicocca, 3rd May, Milan, Italy.

Gusmao Caiado R. G., Filho Leal W., Quelhas O. L. G., Nascimento D. L. M., & Avila, L. V. (2018). A literature-based review on potentials and constraints in the implementation of the sustainable development goals. *Journal of Cleaner Production, 198*, 1276–1288.

Haimes, Y. Y. (2009). On the definition of resilience in systems. *Risk Analysis, 29*(4), 498–501.

Hajer M., Nilsson M., Raworth K., Bakker P., Berkhout F., De Boer Y., ... Kok M. (2015). Beyond Cockpit-ism:

Four Insights to Enhance the Transformative Potential of the Sustainable Development Goals. *Sustainability*, 7(2), 1651–1660

HCSS (2015). Climate Change Vulnerability Monitor. Retrieved from http://projects.hcss.nl/monitor/

Institute of Chartered Accountants in Scotland (ICAS) and International Integrated Reporting Council (IIRC). (2017). New approach for aligning doing business to the Sustainable Development Goals. Retrieved from https://integratedreporting.org/news/new-approach-for-aligning-doing-business-to-the-sustainable-development-goals/

Intergovernmental Panel on Climate Change (IPCC). (2013a). In Stocker, T. F., Qin, D., Plattner, G.-K., Tignor, M., Allen, S. K., Boschung, J., ..., Midgley, P. M. (Eds.), *Climate change 2013: The physical science basis. Contribution of Working Group I to the fifth assessment report of the Intergovernmental Panel on Climate Change* (pp. 1535). Cambridge: Cambridge University Press. Retrieved from http://www.climatechange2013.org/report

Intergovernmental Panel on Climate Change (IPCC). (2013b) *"Climate Change 2013: The Physical Science Basis. Summary for Policymakers"*, Retrieved from http://www.climatechange2013.org/images/report/WG1AR5_SPM_FINAL.pdf

Intergovernmental Panel on Climate Change (IPCC). (2014). *Fifth assessment report – climate change 2014 synthesis report*. Cambridge: Cambridge University Press. Retrieved from http://www.ipcc.ch/report/ar5/

International Council for Science. (2017). A guide to SDG interactions: from science to implementation. Retrieved from https://council.science/publications/a-guide-to-sdg-interactions-from-science-to-implementation/

International Energy Agency. (2013). Resources to reserves – Oil, gas and coal technologies for the energy markets of the future. Retrieved from http://www.iea.org/publications/freepublications/publication/Resources2013.pdf

International Energy Agency. (2015a). World energy outlook special briefing for COP21. Retrieved from www.iea.org/media/news/WEO_INDC_Paper_Final_WEB.PDF

International Energy Agency. (2015b). World energy outlook special report – Energy and climate change. Retrieved from http://www.iea.org/publications/freepublications/publication/WEO2015SpecialReportonEnergyandClimateChange.pdf

International Energy Agency. (2016c). Key CO_2 emission trends – Excerpt from CO_2 emission from fuel combustion. Retrieved from http://www.iea.org/publications/freepublications/publication/KeyCO2EmissionsTrends.pdf

International Integrated Reporting Council (IIRC). (2013). The International (IR) Framework, December. Retrieved from https://integratedreporting.org/wp-content/uploads/2013/12/13-12-08-THE-INTERNATIONAL-IR-FRAMEWORK-2-1.pdf

Jabareen, Y. (2013). Planning the resilient city: Concepts and strategies for coping with climate change and environmental risk. *Cities*, *31*, 220–229. http://dx.doi.org/10.1016/j.cities.2012.05.004

Jusoh, L. S., & Hashim, H. (2018). Development of a framework for greenhouse gas emissions accounting for industry reporting. *Chemical Engineering Transactions*, *33*, 439–444.

Kander, A., Jiborn, M., Moran, D. D., Wiedmann, T. O. (2015). National greenhouse-gas accounting for effective climate policy on international trade. *Natural Climate Change Issue, 5*, 431–435.

Karpoff J., Lott J. R., & Wehrly E. W. (2005). The Reputational Penalties for Environmental Violations: Empirical Evidence. *Journal of Law and Economics*, *48*(2), 653–675.

Kelman, I., & West, J. J. (2009). Climate Change and Small Island Developing States: A Critical Review. *Ecological and Environmental Anthropology*, *5*(1), Retrieved from http://citeseerx.ist.psu.edu/viewdoc/download?doi=10.1.1.611.8076&rep=rep1&type=pdf.

Leal Filho, W., Azeiteiro, U., Alves, F., Pace, P., Mifsud, M., Brandli, L., Caeiro, S. S. & Disterheft, A. (2017). Reinvigorating the sustainable development research agenda: The role of the sustainable development goals (SDG). *International Journal of Sustainable Development & World Ecology*, *25*(2), 131–142. Retrieved from https://www.tandfonline.com/doi/full/10.1080/13504509.2017.1342103

Lenton, T. M. (2011). Early warning of climate tipping points. *Nature Climate Change*, *1*(4), 201–209. http://dx.doi.org/10.1038/nclimate1143

Lettera enciclica Laudato si' del Santo Padre Francesco sulla cura della casa comune, libreria editrice vaticana [Encyclical letter laudato si' by Saint Francis on the care of common house] (2015). Retrieved from http://w2.vatican.va/content/dam/francesco/pdf/encyclicals/documents/papa-francesco_20150524_enciclica-laudato-si_it.pdf

Levitus S., Antonov J. I., Boyer T. P., Locarnini, R. A., Garcia, H. E., & Mishonov A. V. (2009). Global ocean heat content 1955–2008 in light of recently revealed instrumentation problems. *Geophysical Research Letters*, *36*, L07608. https://doi.org/10.1029/2008GL037155

Lewis, J. (2009). An island characteristic: derivative vulnerabilities to indigenous and exogenous hazards. *Shima*:

The International Journal of Research into Island Cultures, 3(1), 3–15.

London Stock Exchange Group. (2017, February). Revealing the full picture. Your guide to ESG reporting. Retrieved from https://www.lseg.com/sites/default/files/content/images/Green_Finance/ESG_Guidance_Report_LSEG.pdf

McGillivray, M., Naude, W., & Santos-Paulino, A. (2008). Achieving growth in the Pacific islands: an introduction. *Pacific economic bulletin*, 23(2), 97–101.

McGray, H., Bradley, R., Hammill, A., with Schipper, E. L. & Parr, J.-E. (2007). Weathering the Storm: Options for Framing Adaptation and Development. World Resources Institute. Retrieved from http://www.wri.org/publication/weathering-the-storm

McKinsey Center for Business and Environment, C40 Cities. (2017). Focused acceleration: A strategic approach to climate action in cities to 2030. Retrieved from https://c40-production-images.s3.amazonaws.com/researches/images/66_MCBE_C40_Focused_Acceleration_report.original.pdf?1510424835

McWilliams A., & Siegel, D. S. (2001). Corporate Social Responsibility: A theory of the firm perspective. *The Academy of Management Review*, 26(1), 117–127.

Mercer, J. (2010). Disaster risk reduction or climate change adaptation: Are we reinventing the wheel? *Policy Arena*, 22(2), 247–264.

Merz T. (2016). *Sustainable Investing Is Becoming Much More Important*, Investment & Pensions Europe (IPE) Magazine. Retrieved from https://www.ipe.com/sustainable-investing-is-becoming-much-more-important/10013232.article

Morgan Stanley Institute for Sustainable Investing. (2015a, March). Sustainable reality: Understanding the performance of sustainable investment strategies. Retrieved from https://www.morganstanley.com/sustainableinvesting/pdf/sustainable-reality.pdf

Morgan Stanley Institute for Sustainable Investing. (2015b, February). Sustainable signals: The individual investor perspective. Retrieved from http://www.morganstanley.com/sustainableinvesting/pdf/Sustainable_Signals.pdf

Mpandeli, S., Naidoo, D., Mabhaudhi, T., Nhemachena, C., Nhamo, L., Liphadzi, S., Hlahla, S., & Modi, A.T. (2018). Climate change adaptation through the water-energy-food nexus in southern Africa. *International Journal of Environmental Research and Public Health*, 15(10), 2306. Retrieved from http://www.mdpi.com/1660-4601/15/10/2306

Nagy, Z., Kassam, A., & Lee, L.-E. (2015, June). Can ESG add alpha? An analysis of ESG tilt and momentum strategies. MSCI. Retrieved from https://www.msci.com/documents/10199/4a05d4d3-b424-40e5-ab01-adf68e99a169

Nilsson, M., Chisholm, E., Griggs, D., Howden-Chapman, P., McCollum, D., Messerli, P., ... Stafford-Smith, M. (2018). Mapping interactions between the sustainable development goals: Lessons learned and ways forward. *Sustainability Science*, *13*, 1489.

Nilsson, M., Griggs, D., & Visbeck, M. (2016). Map the interactions of sustainable development goals. *Nature*, *534*, 320–322.

O'Brien, R., Leichenko, U., Kelkar, H., Venema, G., Aandahl, H., Tompkins, A., et al. (2004). Mapping vulnerability to multiple stressors: Climate change and globalization in India. *Global Environmental Change, 14*, 303–313.

Oak Ridge National Laboratory. (2018). Energy related CO_2 emissions per capita. Retrieved from https://www.ornl.gov/

OECD. (2014, September). Cities and climate change. Policy perspectives. National governments enabling local actions. Retrieved from http://www.oecd.org/

OECD. (2015). *Performance-related budgeting and supreme audit institutions, in Government at a Glance 2015*. Paris: OECD Publishing. Retrieved from https://doi.org/10.1787/gov_glance-2015-29-en

OECD. (2018). *Effective Carbon Rates 2018: Pricing Carbon Emissions through taxes and emissions trading*. Paris: OECD Publishing. Retrieved from https://doi.org/10.1787/9789264305304-en

Orlitzky, M., & Benjamin, J. D., (2001). Corporate social performance and firm risk: A meta-analytic review. *Business & Society*, 40(4), 369–396.

Paris 2015 – UN climate change conference COP21–CMP11. Retrieved from https://unfccc.int/process-and-meetings/conferences/past-conferences/paris-climate-change-conference-november-2015/paris-climate-change-conference-november-2015

Pelling, M., & Uitto, J. I. (2001). Small island developing states: natural disaster vulnerability and global change. *Global Environmental Change Part B: Environmental Hazards*, 3(2), 49–62.

Petit, J. R., et al. (1999). Climate and atmospheric history of the past 420,000 years from the Vostok ice core, Antarctica. *Nature*, 399, 426–436.

Peterson, T. C., & co-authors. (2009). State of the climate in 2008. *Bulletin of the American Meteorological Society*, 90(8), S17–S18 (Special Supplement)

Paavola, J., & Adger, W. N. (2006). Fair adaptation to climate change. *Ecological Economics*, 56(4), 594–609.

Polyak, L., & co-authors. (2009, January). History of sea ice in the Arctic in past climate variability and change in the Arctic and at the high latitudes. The US Geological Survey, The Climate Change Science Program, and the Evaluation Product 1.2.

Pradhan P., Costa L., Rybski D., Lucht W., & Kropp, J. P. (2017). A systematic study of Sustainable Development Goal (SDG) interactions. *Earth's Future*, 5(11), 1169–1179. Retrieved from http://doi.wiley.com/10.1002/2017EF000632

Principles for Responsible Investment (PRI). (2017). *Delivering an ambitious agenda*. Annual Report 2017. Delivering an ambitious agenda. Retrieved from https://www.unpri.org/download?ac=3976

Principles for Responsible Investment (PRI). (2020). About reporting and assessment. Retrieved from https://www.unpri.org/report/about-reporting-and-assessment

Principles for Responsible Investment. (2015). Addressing ESG factors under ERISA. Retrieved from https://www.unpri.org/news/pri-presents-legal-perspectives-on-addressing-esg-factors-under-erisa

Principles for Responsible Investment. (2016). Global guide to responsible investment regulation. Retrieved from https://www.unpri.org/explore/?q=global+guide&hd=on&hg=on&he=on&ptv=&tv=&sp=pub&sc=line&se=start

Ramos, T. B., Martins, I. P. Martinho, A. P., Douglas, C. H., Painho, M., & Caeiro, S., (2014). An open participatory conceptual framework to support State of the Environment and Sustainability Reports. *Journal of Cleaner Production*, 64, 158–172.

Sabine C. L., Feely R. A., Gruber N., Key R. M., Lee K., Bullister J. L., Wanninkhof R., Wong C.S., Wallace D.W.R.,

Tilbrook B., Millero F.J., Peng T.-H., Kozyr A., Ono T., Rios A.F. (2004). The oceanic sink for anthropogenic CO_2. *Science*, *305*, 367–371.

Salvia, A. L., Leal Filho, W., Brandli, L. L., Griebler, J.L. (2019). Assessing research trends related to Sustainable Development Goals: local and global issues. *Journal of Cleaner Production*, *208*, 841–849.

Sands, P. (1992).The United Nations framework convention on climate change. *Review of European Community and International Environmental Law*, *1*(3), 270–277. doi:10.1111/j.1467-9388.1992.tb00046.x

Shalatek, L., Nakhooda, S., Watson, C. (2015). The Green Climate Fund, Climate Funds Update. Retrieved from https://www.odi.org/sites/odi.org.uk/files/odi-assets/publications-opinion-files/10066.pdf

Schaltegger, S., Etxeberria, I. A., & Ortas, E., (2017). Innovating Corporate Accounting and Reporting for Sustainability – Attributes and Challenges. *Sustainable Development*, *25*(2), 113–122.

O'Shea (2003). A review of Gammiella Broth. in Africa, with a range extensionto the East African islands and southern Africa. *Tropical Briology*, *24*, 7–10.

Schipper, E. L. F. (2016). *Climate change adaptation and development: Exploring the linkages*. Norwich: Tyndall Centre for Climate Change Research.

Schoon, I., & Bynner, J. (2003). Risk and resilience in the life course: Implications for interventions and social policies. *Journal of Youth Studies*, *6*(1), 21–31. doi:10.1080/1367626032000068145

Sivakumar, M. V. K. (2005). Impacts of natural disasters in agriculture, rangeland and forestry: An overview. In Sivakumar,

M. V., Motha, R. P., Das, H. P. (Eds.), *Natural disasters and extreme events in agriculture* (pp. 1–22). Berlin: Springer.

Solow, A. R. (1991). Is there a global warming problem? In R. Dornbush & J. M. Poterba (Eds.), *Global Warming: Economic Policy Responses*. Cambridge, MA: The MIT Press Cambridge.

Spina, A. (2015). Reflections on science, technology and risk regulation in Pope Francis' Encyclical Letter Laudato Si'. *European Journal of Risk Regulation*, 6(4), 579–585.

Storey, M., Killian, S., & O' Regan, P., (2017). Responsible management education: Mapping the field in the context of the SDGs. *The International Journal of Management Education*, 15(2), 93–103.

Stymne, S., & Jackson, T. (2000). Intra-generational equity and sustainable welfare: a time series analysis for the UK and Sweden. *Ecological Economics*, 33(2), 219–236.

Sustainability Accounting Standards Board. (2016). *Climate risks. Technical bulletin#: TB001-10182016*. Retrieved from https://www.eenews.net/assets/2016/10/20/document_cw_01.pdf

Sustainable Stock Exchanges Initiative (2016), Model guidance on reporting ESG information to investors. Retrieved from https://sseinitiative.org/wp-content/uploads/2019/12/SSE-Model-Guidance-on-Reporting-ESG.pdf

Task Force on Climate Related Financial Disclosures (TCFD). (2016). Phase I Report of the Task force on climate related disclosures, 31 March 2016. Retrieved from https://www.fsb-tcfd.org/wp-content/uploads/2016/03/Phase_I_Report_v15.pdf

Task Force on Climate Related Financial Disclosures (TCFD). (2017). Final report – Recommendations of the Task force on climate-related financial disclosures, 15 June

2017. Retrieved from https://www.fsb-tcfd.org/wp-content/uploads/2017/06/FINAL-TCFD-Report-062817.pdf

UCLG. (2019). Gold V Report. The localization of the Global Agendas. How local action is trasforming territories and communities. Retrieved from https://www.uclg.org/sites/default/files/goldv_en.pdf

UCLG. (2014). Global Task Force of Local and Regional Developments for post-2015 Development Agenda Towards Habitat III. How to localise targets and indicators of the Post-2015 Agenda. Retrieved from https://www.uclg.org/en/issues/2030-agenda-sustainable-development

UN Comtrade. (2018). International Trade Statistics Database. Retrieved from https://comtrade.un.org/

United Nations Conference on Trade and Development (UNCTAD). (2014). World Investment Report 2014: Investing in the SDGs: an action plan. Retrieved from https://unctad.org/en/PublicationsLibrary/wir2014_overview_en.pdf

United Nations Framework Climate Change (UNFCC). (2014). Focus: Nationally determined contributions (NDCs). Retrieved from http://unfccc.int/focus/items/10240.php

United Nations Development Programme (UNDP). (2016). UNDP Support to the implementation of Sustainable Development Goal 13. Retrieved from http://www.undp.org

United Nations Development Programme (UNDP). (2017). Human Development Report 2016. Retrieved from https://www.undp.org/content/undp/en/home/librarypage/hdr/2016-human-development-report.html

United Nations Environment Programme (UNEP). (2015). *The financial system we need: Aligning the financial system with sustainable development*. Retrieved from http://unepinquiry.org/wp-content/uploads/2015/11/The_Financial_System_We_Need_EN.pdf

United Nations Framework Convention on Climate Change (UNFCCC). (2005). Climate Change: Small Island Developing States. UNFCCC: Bonn.

United Nations Framework Convention on Climate Change (UNFCCC). (2006). Technologies for Adaptation to Climate Change. UNFCCC: Bonn.

United Nations Framework Convention on Climate Change *"The Paris Agreement"* (2015, December). Retrieved from unfccc.int/files/essential_background/convention/application/pdf/english_paris_agreement.pdf

United Nations World Commission on Environment and Development (UNWCED). (1987). Report of the World Commission on Environment and Developmen. *Our common future*. Retrieved from http://netzwerk-n.org/wp-content/uploads/2017/04/0_Brundtland_Report-1987-Our_Common_Future.pdf

UNISDR. (2004). Retrieved from https://www.unisdr.org/2004/campaign/pa-camp04-what-can-you-do-eng.htm

Verner, G., Schütte, S., Knop, J., Sankoh, O., & Sauerborn, R. (2016). Health in climate change research from 1990 to 2014: positive trend, but still underperforming. *Global Health Action*, 9(1). DOI: 10.3402/gha.v9.30723.

Wilhite, D. A. (2000). Drought as a natural hazard: Concepts and definitions. Retrieved from https://digitalcommons.unl.edu/cgi/viewcontent.cgi?article=1068&context=droughtfacpub

World Business Council for Sustainable Development. (2017, January 18). Sustainability and enterprise risk management:

The first step towards integration. Retrieved from www.wbcsd.org/contentwbc/download/2548/31131

World Economic Forum. (2016). The global risks report. Retrieved from http://www3. weforum.org/docs/GRR/WEF_GRR16.pdf

FURTHER READINGS

2° Investing Initiative. (2015a). *Assessing the alignment of portfolios with climate goals: Climate scenarios translated into a 2°C benchmark*. Working Paper. Retrieved from http://2degrees-investing.org/IMG/pdf/2dportfolio_v0_small.pdf

2° Investing Initiative. (2015b). *Financial risk and the transition to a low-carbon economy: Towards a carbon stresstesting framework*. Working Paper realized in partnership with the UNEP Inquiry and CDC Climate Research, Retrieved from http://www.2degrees-investing.org/#!/page_Resources

Arabella Advisors. (2016, December). The global fossil fuel divestment and clean energy investment movement. Retrieved from https://www.arabellaadvisors.com/wp-content/uploads/2016/12/Global_Divestment Report_2016.pdf

ASviS. (2017). Festival dello Sviluppo Sostenibile *2017* – Disegniamo il futuro [Meeting of Sustainable Development 2017 - Writing the future]. Cambiamo il presente [Changing the present]. Calendario Eventi [Schedule events], 22 May–7 June. Retrieved from https://festivalsvilupposostenibile.it/2017

Bank of England, Arthur Burns Memorial Lecture, Berlin (2016, September 22). Resolving the climate paradox.

Speech given by Mark Carney Governor of Bank of England, Chairman of the Financial Stability Board. Retrieved from http://www.bankofengland.co.uk/publications/Documents/speeches/2016/speech923.pdf

Bank of England, Loyds of London. (2015, September 29). Breaking the tragedy of the horizon – Climate change and the financial stability. Speech given by Mark Carney Governor of Bank of England, Chairman of the Financial Stability Board. Retrieved from http://www.bankofengland.co.uk/publications/Documents/speeches/2015/speech844.pdf

Briand, R., Lee, L., Lieblich, L., Menou, V., & Singh, A. (2015, April). *Beyond divestment: Using low carbon indexes*. MSCI. Retrieved from https://www.msci.com/documents/10199/031bf397-5920-4fef-b743-0c879ae46610

Canfin-GranJean Commission. (2015, June). Mobilizing climate finance – A roadmap to finance a low carbon economy. Retrieved from http://www.cdcclimat.com/IMG/pdf/exsum-report_canfin-grandjean_eng.pdf

Carbon Trust (2008). Climate change – A business revolution? How tackling climate change could create or destroy company value". Co-analysis with Mc Kinsey & Co. Retrieved from www.carbontrust.com/media/84956/ctc740-climate-change-a-business-revolution.pdf

Climate Council. (2015). Climate change 2015: Growing risks, critical choices. Retrieved from https://www.climatecouncil.org.au/uploads/153781bfef5afe50eb6adf77e650cc71.pdf

EU High-Level Expert Group on Sustainable Finance. (2017, July). *Financing a sustainable european economy*. Interim Report. Retrieved from https://ec.europa.eu/info/sites/info/files/170713-sustainable-finance-report_en.pdf

Global Sustainable Investment Alliance (GSIA). (2016). Trends report 2016. Retrieved from http://www.gsi-alliance.org/members-resources/trends-report-2016/

Mckibben, B. (2012, July). Global warming's terrifying new math: Three simple numbers that add up to global catastrophe – And that make clear who the real enemy is. Rolling Stone. Retrieved from http://www.rollingstone.com/politics/news/global-warmings-terrifying-new-math-20120719

World Economic Forum. (2016). The global risks report. Retrieved from http://www3.weforum.org/docs/GRR/WEF_GRR16.pdf

INDEX

Accountability system, 39
Adaptation, 42
   climate change, 29
Agenda 2030 (*see*
   2030 Agenda
   for Sustainable
   Development)
Agriculture, 34
Alarming, 9
Anthropogenic GHG
   emissions, 50
Arctic sea ice, 48

Benchmarking methods, 60
Best in class selection, 66
Budgeting practices
   in executive, 38
   and procedures, 40
Building Resilient in Power
   Sector, 82
Business Theme GHG
   Emission, 80–82
Businesses, 85
   context, 46–47
   leaders, 73

C40 Cities Finance
   Facility, 44
Capacity-building for
   climate change, 26
Carbon dioxide (CO_2), 32
Carbon emissions, 79
Casablanca Finance City
   Authority, 71
Casablanca statement on
   financial centers for
   sustainability, 71
Cities Development
   Initiative for
   Asia, 44
City Inventory Reporting
   and Information
   System tool, 44
Climate action for URBan
   sustainability
   (CURB), 44
Climate Action Tracker
   (CAT), 38
Climate change, 3–4, 7–8,
   34, 83
   adaptation, 29
   build knowledge and
      capacity to meet,
      26–30
   climate change-related
      effects, 8

education on, 26
integrating climate change measures into policy and planning, 25–26
promoting mechanisms to raising capacity for planning and management, 27–30
SDG 13 and main focus on, 21–22
urgent action to combating climate change and impacts, 23
Climate Change Agreement, 31–32
Climate risk assessment, 47–56
Climate Transparency, 38
Climate-related disasters, 23–25
CO_2 equivalent (CO_2-eq), 51
Colour-based method, 15
Conflict, 28
Contextualisation of climate actions, 30
Coordinating units in executive, 38
COP21, 22, 31, 42
COP22, 9, 42
Corporate reporting, new trends in, 73–74

Direct emissions, 31
Displacement, 24, 28

Education on climate change, 26
Empowerment, 7
Engagement and active ownership, 66
Environmental, social and governance (ESG), 47, 56, 58, 61–68
factors, 75
integration, 68
issues, 62–64
Environmental reporting, 75
European Fund and Asset Management Association (EFAMA), 47
European pension funds, 75
European Union (EU), 39
Eurostat, 39
Exclusion process, 65–66
Exclusionary screening, 65
Extreme events, 49

Financial Centers for Sustainability, 70–71
Financial Stability Board (FSB), 54–55
Financial system, 70
Fluorinated gases, 32
Fossil fuels, 5
Fragility, 28

G20 countries, 39
Glaciers, 49
Global Climate Action Summit, 43

Index

Global Covenant of Mayors for Climate and Energy, 41
Global Protocol for Community-scale standard (GPC standard), 44
Global Reporting Initiative (GRI), 77–82
Global sustainable development goals, 3
  actions to implementing SDGs, 10–11
  calculation of SDG indicator at 'country' level, 12–19
  implementation of Agenda 2030 and SDGs, 7–10
Global temperature, 48
Global warming, 37
Governance, 7
Governments for SDG 13, 37–40
Green and sustainable finance, 56
  mobilising, 69–71
Green Climate Fund (GCF), 26–27
Green House Protocol, 79
Greenhouse gas (GHG), 3
Greenhouse gas emissions (GHG emissions), 31, 37, 42
  anthropogenic, 50
  in IPCC scenarios, 51
  RCPs, 51
GRI Report Index, 78–79

High-level Expert Group on sustainable finance (HLEG), 57
High-power gas global warming (Gas high GWP), 32
Homelessness, 24
Human and natural systems, 48
Hydrofluorocarbons, 32

Ice caps, 48
ICLEI's Transformative Actions Programme, 44
Impact investing, 68
Impact Reporting and Investment Standards, 82
Inclusiveness, 45–46
Indicators of SDG 13, 22–30
Indirect emissions, 31
Innovative financial mechanisms, 56–61
Integrated Reporting (IR), 82–83
Integration of climate change into national policies, 25–26
Intended nationally determined contributions (INDCs), 8
Intergovernmental Panel on Climate Change (IPCC), 3, 50
International Council for Science, 33

International Organisation of Supreme Audit Institutions, 40
International policymakers, 8
Investment policy, principles of, 75
Investors, 73

Legal risks, 52
Lesbian, Gay, Bisexual and Transgender (LGBT), 65
Local disaster risk management, 25
Localisation of SDGs, 40–45
Low-carbon benchmark, 60–61

Market risk, 52, 54
McKinsey Centre for Business and Environment, 44
Methane (CH_4), 32
Migration, 28
Millennium Development Goals (MDGs), 7–8
Morgan Stanley Capital International (MSCI), 7
Multi-stakeholder platform, 59

National disaster risk management, 25
National economies, 21
National policymakers, 8
National strategy and baseline assessments in executive, 37–38
Nationally determined contributions (NDCs), 9
Natural capital, 7
Nitric oxide (N_2O), 32
Nitrogen trifluoride, 32
Non-financial information, 74–75
Non-financial statement, 76
Norm-based screening, 65

Ocean acidification, 49
Ocean warming, 48
One Planet Summit, 42
Open participatory approach, 46
Organisation for Economic Co-operation and Development (OECD), 10–11

Paris Agreement, 9, 22, 44, 70
Paris Climate Agreement, 42, 61
Participatory approach, 45–46
Perfluorocarbons, 32
Performance budgeting, 40
Policy frameworks, 43–44
Political risks, 52
Positioning of communication, 55

Positive carbon impact benchmark, 61
Principle of materiality, 55
Private sector, 46–47
Public sector for SDG 13, 37–40

Renewable energy, 5
  mitigation and investments in, 28
Representative concentration paths (RCPs), 51
RCP2. 6, 52
Reputational risk, 54
Rio Earth Summit, 9

Scaled-up climate finance, 29
Scenario analysis, 55–56
Scoring approach, 33
SDG 13, 21, 73
  Climate Change Agreement, 31–32
  ESG and sustainable finance, 61–68
  and focus on climate change, 21–22
  governments and public sector, 37–40
  inclusiveness and participatory approach, 45–46
  innovative financial mechanisms, 56–61
  interactions frameworks, 32–35
  local agenda and key interactions frameworks, 31
  mobilising green and sustainable finance, 69–71
  new trends in corporate reporting, 73–74
  planning, 47–56
  practical tools and mechanisms, 37
  private sector and business context, 46–47
  targets and indicators, 22–30
  voluntary to compulsory disclosures, 74–76
Sea level rise, 21, 48
SGDs Compass Guide, 77–78
Small island developing state (SIDS), 9
Snow cover reduction, 49
Socially responsible investments (SRI), 47
Spending reviews, 40
Stakeholders, 73–75
'Strategic-level' analysis, 44
Sub-national governments, 45
Sulphur hexafluoride, 32
Supreme audit institutions (SAIs), 39
Sustainable agricultural practices, 34
Sustainable development, 6, 69

Index

Sustainable Development Goals (SDGs), 1, 6–7
  actions to implementing, 10–11
  calculation of SDG indicator, 12–19
  and GRI, 77–82
  implementation, 7–10, 86
  and IR, 82–83
  localisation of SDGs, 40–45
  most and least-researched globally, 87
  SDG 7, 35
  SDG 14, 35
  SDG Index & Dashboards, 12, 18–19
Sustainable Development Solutions Network (SDSN), 12
Sustainable finance, 61, 64–68
Sustainable investment, 64

Targets, 6
  of SDG 13, 22–30
Task Force on Climate-related Financial Disclosures (TCFD), 54
Technological risk, 54
Thematic investment, 67
Transformative action in food systems, 34

2030 Agenda for Sustainable Development, 4, 28, 44–45, 70, 82
  implementation, 7–10

UCLG, 41–42
UN Framework Convention on Climate Change (UNFCCC), 8
UNDP, 30
  Human Development Index, 14
United Nations Environment Programme (UNEP), 69
United Nations Framework Convention on Climate Change, 26–27
United Nations Global Compact (UNGC), 77
Urban-climate nexus, 43

Voluntary to compulsory disclosures, 74–76

Well-being of Future Generations Act, 43
World Business Council for Sustainable Development (WBCSD), 77
World Federation of Exchanges (WFE), 61

Zero-carbon development, 29